System Administration

McGraw-Hill Titles on LOTUS NOTES/DOMINO

KELLEHER, Rose, et. al.: ***Advanced Domino 5 Web Programming.***
0-07-913691-5

LAMB, John; LEW, Peter: ***Lotus Notes & Domino Network Design.***
0-07-913241-3

OLIVER, Steve: ***Accelerated Domino Web Development and
Administration Study Guide.*** 0-07-134533-7

OLIVER, Steve; WOOD, Pete: ***Lotus Domino Web Site Development.*** 0-07-913755-5

SCHWARZ, Libby Ingrassia: ***Accelerated Lotus Notes Application
Development Study Guide.*** 0-07-134569-8

SCHWARZ, Libby Ingrassia: ***Accelerated Lotus Notes System
Administration Study Guide.*** 0-07-134562-0

THOMAS, Scott; HOYT, Brad: ***Lotus Notes & Domino Architecture
Administration and Security.*** 0-07-064562-0

THOMAS, Scott; PEASLEY, Amy: ***Lotus Notes Certification Exam
Guide: Application Development and System Administration.*** 0-07-913674-5

THOMPSON, William: ***Accelerated LotusScript Study Guide.***
0-07-134561-2

System Administration

Accelerated Lotus Notes Study Guide

Libby Ingrassia Schwarz

McGraw-Hill
New York • San Francisco • Washington, D.C. • Auckland
Bogotá • Caracas • Lisbon • London • Madrid • Mexico City
Milan • Montreal • New Delhi • San Juan • Singapore
Sydney • Tokyo • Toronto

McGraw-Hill

A Division of The McGraw·Hill Companies

Copyright © 1999 by The McGraw-Hill Companies, Inc. All rights reserved.
Printed in the United States of America. Except as permitted under the United
States Copyright Act of 1976, no part of this publication may be reproduced or
distributed in any form or by any means, or stored in a database or retrieval system, without the prior written permission of the publisher.

1 2 3 4 5 6 7 8 9 0   AGM/AGM   9 0 3 2 1 0 9 8

ISBN 0-07-134562-0

The sponsoring editor for this book was Judy Brief and the production
supervisor was Sherri Souffrance. It was set in Stone by D & G Limited, LLC.

Printed and bound by Quebecor/Martinsburg.

McGraw-Hill books are available at special quantity discounts to use as
premiums and sales promotions, or for use in corporate training programs.
For more information, please write to Director of Special Sales, McGraw-Hill,
11 West 19th Street, New York, NY 10011. Or contact your local bookstore.

This book is printed on recycled, acid-free paper containing a minimum of 50% recycled de-inked fiber.

Dedication

This book is dedicated to my grandparents. You taught me about work, loyalty, love, and pride in myself. Thank you.

And, always, to my wonderful husband Brian, just for being.

Acknowledgments

Many people deserve thanks and acknowledgment for making it possible for me to write this book.

First, many thanks to my Total Seminars Houston "family": Dudley, Mike, Scott, Cassandra, Janelle, Bob, and Roger.

Second, a sigh of appreciation to all my friends and family who understood each time I said "I can't, I'm working . . . ."

Finally, to all my Notes colleagues at DRT, my technical writing colleagues, and my training colleagues . . . thank you for helping me learn how to do all of this!

Contents

Introduction to Lotus Notes

When you purchase Lotus Notes books or tell people that you are studying for a Lotus Notes certification examination, the first question you will hear is, "So what exactly *is* Notes, anyway? Some sort of mail client, right?" This is when you begin your sideline career explaining Lotus Notes and the concept of groupware.

Lotus Notes markets itself as the premier groupware product on the market, competing with Microsoft Exchange, Novell GroupWise, and others. *Groupware* is any software that enhances the abilities of groups of people (such as departments and workgroups) to communicate and do business. Groupware usually has multiple functions, including e-mail, knowledge management, and collaboration. Although Lotus Notes functions well as strictly an e-mail engine, Notes can do much more. Notes can also perform a variety of other workgroup communication functions, including sharing documents, automating business processes using workflow applications, publishing on the World Wide Web, creating internal corporate intranets and more.

Often, it is easier to define Notes by describing how it can function in an office environment. Here are two examples. First, Notes works much like discussion groups on the Internet. In a discussion group, you have the ability to write a message containing information and share it with multiple people, without having to e-mail the message to everyone. This is useful for two reasons. First, having the capacity to share information without using e-mail can save an amazing amount of space on your server(s). Second, when people leave or join an organization or workgroup, there is the potential for either information to be lost or not to be passed on efficiently. It is easier to give a new user in a workgroup rights to a database that contains all the shared data on a project, rather than to forward all the e-mail related to the project.

Lotus Notes can reduce both the amount of paper used in an office and the time it takes to finalize a request through multiple people. One example I give as an instructor when trying to describe Notes is to ask one of my students to describe a business process in their organization—such as asking for time off or requesting expense reimbursements. The student usually describes a process of filling out a form and then walking or sending it to the appropriate person for approval. That person would then need to forward the form to three or four more people before the process was complete. This often included multiple people with various places the form could be sidetracked or lost—often without any real way of recovery. Notes can automate such processes, allowing a person to track the progress of a request, find out who signed it and who did not, and even send a memo to the latter to nudge them into action. A Notes developer can customize Notes to accomplish just about any task done in an office setting. In the previous example, a Lotus Notes developer could create the request form in a database. Once the database is completed and saved as a document, the developer would place that form in a view in the database—where the approval process will take place. This leaves an electronic paper trail for all those involved in the process to track the status and location of the form. As a Notes System Administrator, you probably will not create databases directly to fulfill the business needs of your organization. That is generally the duty of Notes Application Developers. You will most likely, however, create the Notes environment and servers in which those databases exist. You will also have to troubleshoot these applications and provide support to application users. Finally, you will provide security, mail, and other Notes default functions for the environment. As a candidate for certification, you are expected not only to create an environment in which Notes applications can exist and be used, but also to provide the other basic functions of Notes that make the applications work in expected ways.

With the release of Notes 4.5, Lotus introduced the term *Domino* to describe the server side of Notes, while reserving the term *Notes* to describe the workstation side of the software. Prior versions of Lotus Notes used *Notes* to describe both server and workstation elements of the software. The terminology is less confusing now, but unfortunately for you, the System Administration exams were written *before* R4.5 and therefore use the older, more confusing terms. To make this even more fun, you will probably prepare for the exams by using a later version of Notes, either 4.5X or 4.6X. In these versions, just remember that the term Domino refers to the server portion of the software.

Introduction to this Book

Certifications by software and hardware vendors are more than just popular in the information technology industry—they have become necessary. These certifications enable companies to hire employees or consultants who have proven their knowledge of the products they will implement or support.

Like many other vendors, Lotus implemented a set of vendor certifications for Lotus Notes. To obtain these certifications, you will need knowledge of the Domino and Notes environments, and you will need to prove this knowledge by taking the Lotus Education exams. These certifications are valuable for many reasons. First, salary surveys in a variety of magazines consistently show higher salaries for certified employees. For those who are trying to find their niche in the busy information technology industry, too, a certification is often necessary to receive that all-important first interview or job. Everyone has a different method of preparing for certification exams. Some take the official Lotus Education courses offered by Lotus authorized education centers. Others attempt to study on their own or rely on their knowledge of the software. Others look at the local bookstore for a guide to passing the exams. While there are many good books on the market to assist in designing and implementing a Notes application or environment, there are relatively few Notes certification study guides.

This book is a Notes certification study guide. It is intended to be a quick reference for the information covered on the System Administration I and II exams. No matter what other types of studying you choose, this book will help you refine your study of the material actually covered on the exams.

The goal of this book, therefore, is to help you to pass the exams necessary to become certified in Lotus Notes system administration. Specifically, the book is intended to prepare you for passing the multiple-choice System Administration I and System Administration II exams offered by Lotus through its independent testing vendors.

To that end, each section of the book has objectives relating to questions at the end of each section. These objectives are based on the lists of competencies published by Lotus Education in their certification exam guides. They are also based on personal experiences taking the Lotus exams, teaching the official Lotus courses, and implementing Notes enterprise environments for a variety of companies.

To help you study, I recommend reading each chapter first, looking for information required to understand the objectives. Then, you should answer the questions at the end of each section immediately after reading that particular section. Later in your studying, go through the chapters again. This time, read the bulleted lists and definitions. You also will want to look over the screen captures and graphics again. Then, go over the questions one more time to review concepts and make sure you have not forgotten anything. You will notice that certain concepts and details are covered more than once in the text, and this is not an insult to your intelligence. It is one way of ensuring that you have the opportunity to review important topics several times in the course of your studying. The topics that you see multiple times in the text are covered in depth in the courses and official certification guides. You should expect to see them covered thoroughly on the exams.

In addition to reading this book and answering the practice questions in each section, I would highly recommend that you work with Lotus Notes. For most of the exercises in System Administration I, you need access to both a Notes client and server. You also will need a user ID that has a regular Notes license. When you decide to set up a server to create your ID and to work through the server topics, keep in mind that you do not need any fancy hardware to set up a Notes server. It can run on almost any operating system. If you do not already work with Notes, you can obtain a trial version through the Internet from the Notes Net at http://notes.net. The exams assume that you have experience with all of the basic administration tasks, including planning and installing a Notes environment. The Notes environment should have more than one server, if possible. You will need to register users and servers and implement the basic Notes functions of e-mail, security, and replication. You cannot get this experience from any other source. This book assumes, therefore, that you have some experience with Lotus Notes—at least to have some familiarity with the functions and feel of the software.

Lotus Certification Paths and Resources

Certifications awarded by Lotus for Notes R4.X include the *Certified Lotus Specialist* (CLS), the *Certified Lotus Professional* (CLP), the *Principal Certified Lotus Professional* (Principal CLP), Domino Messaging Administrator (R4.6 only), and the *Certified Lotus Instructor* (CLI). The exams covered in this book are required for CLP, Principal CLP, and CLI for Notes R4 to 4.5. For Notes R4.6 and higher, additional exams and certifications have been released by Lotus. Read on for more information. The System Administration I exam is one of the options for CLS status. When you combine System Administration I and II with a Application Development I exam (covered in another McGraw-Hill accelerated guide), you will be a Certified Lotus Professional—System Administrator.

Becoming a Certified Lotus Specialist requires you to pass only one examination. Exams required for CLS are displayed in the following exam matrix table. To become a CLP, however, you must take three exams. The two basic, required exams are System Administration I and Application Development I. For the third exam, the certification candidate gets to choose whether to specialize in System Administration or Application Development. According to Lotus statistics, approximately 60 percent choose to become Application Developers, and 40 percent choose to become System Administrators.

NOTE
The following exam matrix contains only some of the possible examination paths. Other exams exist, such as the Domino Web Development and Administration, the Developing LotusScript Applications for SmartSuite, and the cc:Mail exams that qualify a candidate for CLS certification. In addition, there are new tracks available for Notes r4.6 certifications, including the Domino Messaging Administrator and the alternative exams available for CLP certifications. For updated information on these alternatives, please see the Lotus Education and Certification Web site at the following URL:

http://www.lotus.com/education

Exam Matrix

Exams for CLS Status (4.5)	Add these exams for CLP Application Developer (4.5)	Add these exams for Principal CLP Application Developer (4.5)	Add these exams for CLP System Administrator (4.5)	Add these exams for Principal CLP System Administrator (4.5)
Application Development I (190–271)	System Administration I (190–274); Application Development II (190–272)	LotusScript in Notes for Advanced Developers (LND) (190–273); Developing Domino Applications for the Web (DDA) (190–278); Domino Web Development and Administration (DWDA) (190–281); Developing LotusScript for SmartSuite 1-2-3 Apps (DLAS) (190–291)	System Administration I (190–274); System Administration II (190–275)	Administrating Specialized Tasks for Domino 4.5 (AST)(190–276); cc:Mail R6 System Administration 1 (190–251)
System Administration I (190–274)	Application Development I (190–271); Application Development II (190–272)		Application Development I (190–271); System Administration II (190–275)	

Getting More Information

Like all Web-based information printed in books, the Lotus exams and certifications described earlier may have changed somewhat by the time you read this. To ensure you have the most updated information, refer to the following sources as you begin your certification path—and as you continue progressing.

- Lotus Education Web page: Lotus' Web site, at http://www.lotus.com, is a rich source of additional information about exams, exam topics and competencies, and the certifications. Specifically, if you go to the education home page at http://www.lotus.com/education, you can find updated information on exams and the exam matrix, as well as updates on the certification paths. You can also find the official Lotus exam guides, which you can download in Adobe Acrobat file format (PDF).

- Lotus Education Help Line: If you have specific questions about taking the examsor taking official Lotus courses, or about the Lotus certifications, you can call the Lotus Education Helpline at (800) 346-6409 or (617) 693-4436. The Helpline is open from 8:30 A.M. to 5:30 P.M. ET, Monday through Friday.

The Exams

After you determine the path for your Lotus Notes certifications, it is time to study for your first exam. If you purchased this book, I assume that you have made some of these decisions, such as which exams to take. You must also choose which testing format you prefer: multiple-choice or concurrent application.

Lotus has two types of examination formats available for the level one exams. For all Lotus exams, there is the typical multiple-choice examination. For both System Administration I and Application Development II exams, however, Lotus also offers concurrent application testing. This type of testing offers the person taking the exam an opportunity to work with Lotus Notes and Domino to accomplish specific tasks during the time allotted. Lotus intends to offer the concurrent application format for the Application Development II and System Administration II exams by the end of 1998.

The multiple-choice tests require the test-taker to answer approximately 40 to 60 questions. These questions are presented in multiple-choice format, including true or false questions, questions with multiple answers, and questions that require users to interpret graphics or diagrams. Each test is computer-based and timed, with the time limit set specifically for each exam. The testing proctor will ensure that you know the time limit, and the test interface will show you how much time you have left during the test. Each test is considered closed-book. You will be given blank scratch paper (which must be returned to the testing proctor after the exam) and a pen or pencil if needed. No other items may be brought into the testing center, including purses, pagers, notes, books, and calculators. At the beginning of the exam, the testing interface will display the exact number of questions and the passing percentage. When you have completed the test, the interface will display your percent correct and the passing percentage. This will also be printed at the testing center for you to keep and will be forwarded to Lotus Education within five business days. If you do not pass one of the exams, you must register and pay the exam fee again through Sylvan Prometric or CATGlobal.

> **NOTE**
> Testing rules of each center are explained to the candidate both at registration time and at testing time. Be aware that Lotus and the testing centers take these rules seriously. If a candidate is accused of and is proven to have passed an exam through any inappropriate or questionable means, the candidate's scores will be dismissed and the candidate will have to retest after a six-month waiting period. The candidate cannot take any Lotus certification exams within the six-month waiting period.

This book specifically targets the multiple-choice-type examination. The information and competencies covered, however, apply to both the multiple-choice and the concurrent application testing formats. Working through the examples and tasks described in each chapter, as well as the review questions, should prepare you for the concurrent application testing format—if that is the format you prefer. More hands-on work with Notes is highly recommended for testing using the concurrent application method.

The next step in pursuing a Notes certification is to study for and take the exams you have selected. Obviously, everyone learns and studies differently. This book is an excellent study tool for all types of learning methods, because I have tried to give you the opportunity to read about the topics that you need to know and to work through some of the tasks that are required competencies. In addition to reading this book and working through the server administration tasks and review questions, you may find that classes such as Lotus' Learning Bytes, computer-based training, and a variety of practice exams will help you pass the exams.

Many find that hands-on experience is the best way to learn software and pass exams. One way to get this experience is to take classes in Notes. Two sets of classes are available to help you pass the Notes certification exams. First, there is the official Lotus curriculum. These are hands-on, instructor-led courses that are usually two to five days long for the technical courses. The end user courses are recommended prerequisites, and each one takes a half-day. You will find many of the same topics from this book covered in the courses. Note, however, that Lotus' stated objective for students in the classes is learning the tasks related to their jobs; i.e., learning to be good system administrators. The courses differ slightly from the exams, in my experience. The official courseware, in addition, has been updated more recently than the exams. The other instructor-led classes available include a variety of unofficial courses. In addition to the instructor-led classes, there are a variety of *computer-based training* (CBT) alternatives. These include the Lotus Learning Bytes, available at `http://notes.net/lbytes.nsf/`, and CD-ROM-based CBT—again, from multiple companies. After using all these methods to study, you may also want to buy some of the practice exams offered on the market. The practice exams not only give you the ability to test your knowledge but also enable you to practice the act of taking an exam, including time limits and computerized testing. Mastering these skills can be a huge help in passing the exams. Finally, as mentioned earlier, you can obtain a trial version of Lotus Notes and Domino from `http://notes.net`. This gives you the chance to practice all the skills and techniques necessary to pass the exams.

How Much Should I Study?

After helping many students through the System Administration I and II exams, I have developed an idea of how long it takes students to master the material and pass the exams. The time it takes to study for the exams is described in the following table. The number of hours listed in the table only represent averages. If you are a slower, more methodical student, or if you are a nervous tester, add 10 percent. If, on the other hand, you are the student who mastered chemistry in one night, you may want to reduce the numbers by 10 percent. To use the table, circle the values under the column which represent how much time you have spent doing the tasks listed. Then add up the values to get a total number of study hours.

- If you are new to Notes in general, add 20 hours to the total.
- If you are comfortable with Notes as an end user or application developer, add 10 hours to the total.
- If you have been a Notes administrator, subtract 10 hours from the total.
- If you are an administrator of another type of environment, but you are not familiar with Notes, you should add 10 hours to the total.

Amount of experience doing . . .	None	Once or Twice	Occasionally	Often
Using the Notes client	10	8	7	5
Using the Notes server console	10	8	7	5
Planning Notes environments	10	8	7	5
Installing and configuring Notes servers	8	6	4	2
Installing and configuring Notes workstations	6	4	2	1
Registering Notes users or servers	8	6	4	2
Using Notes Mail as an end user	8	6	4	2
Configuring Notes Mail	10	8	6	5
Enabling or maintaining shared mail	10	8	6	5

Amount of experience doing . . .	None	Once or Twice	Occasionally	Often
Creating Connection documents	5	4	2	1
Creating Location documents	5	4	2	1
Using Server Replication	10	8	6	5
Using Workstation Replication	6	4	2	1
Renaming or recertifying users	8	5	4	2
Creating and modifying ACLs	8	5	4	2
Using the Public NAB	10	8	6	5
Using or configuring Passthru	5	4	2	1
Creating Cross-Certificates	8	5	4	2
Creating Server security	5	4	2	1
Sending Notes mail outside your Notes domain	8	5	4	2
Configuring mail among multiple domains	8	5	4	2
Monitoring your Notes log	8	5	4	2
Using statistics and events to monitor Notes	10	8	6	5

How to Take the Tests

The following paragraphs outline some test-taking strategies that often make the multiple-choice-style exams easier.

- Use the time before you actually begin the test to jot down any memorized notes on the scratch paper provided.

- Go through the exam once, answering any questions of which you are certain. Make notes of any questions giving you information that might help you answer another question.

- Mark any questions you are not sure of and return to them on your second pass.

- Answer *all* the questions. Blank questions are automatically considered wrong.

- Sometimes it helps to write or draw part of the information given in a question. This helps clarify scenario questions.

- Read each question slowly and carefully. Make sure you know what the question is asking before you answer. It is easy to go too quickly and miss a key word.

- Visualize the Lotus Notes interface as clearly as you can, including menus and dialog boxes. Draw diagrams to help you visualize, if that is necessary. Knowing where options are located and knowing how to perform actions will be important.

Some of these strategies may work for you, while others may not. I recommend using practice exams to try out these ideas to see which work for you. The more at ease you are with the testing format, the more you can concentrate on the questions. Good luck.

REGISTERING FOR THE EXAMS

After completing your studying, you can register for the exams by contacting one of Lotus' two independent testing vendors. For the multiple-choice exams, register by contacting Sylvan Prometric testing centers. To contact Sylvan, use one of the following numbers:

(800) 74-LOTUS, which is (800) 745-6887, or (612) 896-7000.

You may also register online for exams at Sylvan by going to Sylvan Prometric's online registration site at the following URL:

http://www.slspro.com

You will need the following information ready, either on the phone or online, when you register:

- Name
- Social Security number or testing ID
- Mailing address and phone number
- Company name
- Name and number of the exam you are registering to take
- When and where you want to test. You can find testing center information at the Sylvan Prometric Web site: http://www.sylvanprometric.com/nav.htm).
- Payment method (credit card, voucher, money order, or check. Note that you cannot test until they receive payment, which can postpone testing if paying by money order or by check.

To register for concurrent application testing, contact CATGlobal Testing Centers. Complete your CATGlobal registration online at the following URL:

http://www.catglobal.com

CHAPTER 1

Installation and Configuration

One of the competencies tested on the System Administration 1 exam is the ability to plan, install, and configure Notes servers and Notes workstations in your environment. This chapter examines the issues surrounding the architecture, installation, and configuration of a Notes environment. This topic includes naming, Notes Domains, the Public Name and Address book, and other important decisions required during configuration, in addition to the actual process of installation.

NOTE
During the discussion of installation and configuration, you may want to install and configure your own Domino server. This will assist you with practicing the areas covered in this chapter, and it will help you during the rest of your studying for these exams.

Objectives

After reading this chapter, you should be able to answer questions based on the following objectives:

- Plan and implement a Notes Naming Scheme, including Hierarchical Names, Organization and Organizational units, Domains, and Notes Named Networks.

- Plan a Notes Named Network.
- Install and configure a First Server.
- Register additional servers and users.
- Install and configure additional servers.
- Break down a server or workstation.
- Install and configure a workstation.

The Notes Software

Lotus Notes can be divided into two distinct types of software: the Server and the Workstation. The server software, as its name implies, is the serving component. The server software is character-based and runs perfectly well in a command prompt under all versions of Windows. The server software also runs on NetWare, UNIX, OS/2, and others. The server has a prompt and takes command-line input, similar to other text-based operating systems, such as DOS. The workstation is a graphical program and is the primary user interface for Notes. Even though the server can take commands directly from console prompts, 99 percent of all commands to the server can be and are handled through the workstation. A copy of the workstation software will also reside on the server itself. Installing Notes requires installing the client and server software on one or more systems, depending on the needs of the installation, followed by installation of the workstation software on the client computers. Typically, Lotus Notes provides both the server and workstation software on the same CD-ROM.

Planning the Notes Installation

Before installing a Notes server, three important organizational issues need to be addressed. First, you must design the *organization*. An organization is a hierarchical naming scheme applied to every server and user in a Notes environment. The organization grants a unique identifying name to every server and user. Second, you must structure the *domain*. The domain defines pathways for e-mail, a central repository for addresses and configurations, and other administrative features. All these are essential for the proper operation of the mail-based Lotus Notes. The *Public Name and Address Book* (NAB) defines the domain for an environment. (The NAB is discussed more fully in Chapter 2.) The third organizational issue to address is whether to create one or several *Notes Named Networks* (NNN). Organizations are never simple—there can be separate offices and different types of networks that prevent the smooth transport of e-mail. These problems are handled by a subdivision of the Notes Domain—the NNN. A domain can have many NNNs, each handling a subgroup of the domain on a network that uses the same protocol and is constantly connected. If a user sends an e-mail within the NNN, it is automatically and immediately sent to its recipient. If the e-mail is going to a user outside the NNN, Notes checks for a connection document within the NAB to determine how it should be sent. Notes Mail is described in detail in Chapter 4, "Notes Mail". A plan of a sample Notes environment, including information about the organization, domain, and NNNs, is shown in Figure 1.1.

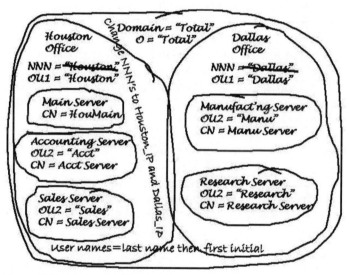

Figure 1.1 Plan for a Notes environment

An important distinction between organizations, domains, and NNNs occurs at installation time. Once a naming convention has been started, it cannot be changed without the massive task of recreating all the names on the network. The same is also true for domains, because you must create a new first server in your organization to create a new domain. NNNs, however, can be created at any time.

In the next three sections, we will gain an understanding of organizations, domains, and Notes Named Networks. We then will begin the installation of the Notes server with a better appreciation of the need for these items.

Planning an Organization

Each organization needs a well-defined naming structure that applies unique names to each server and user in the environment. Naming is an integral part of verifying the rights of users and processes. When a user requests to look at an e-mail, a database, or a document, how does the system know that person has the right to perform the function they requested? Are they allowed to see a particular piece of information? Can they read a certain e-mail? For that matter, where exactly are they? The organization addresses all of these issues. The following discussion covers four basic topics: authentication, flat versus hierarchical naming schemes, the levels in the hierarchical naming scheme, and some terminology specific to hierarchical naming in Notes.

Authentication is the process of verifying the rights of a user or a server to access a server. Authentication is manifested through a series of encoded files with the ID extension. The core identifier, the ID file from which all other ID files are created, is known as the CERT.ID. A new first Notes server installation usually creates the new CERT.ID file. Each level of an organization adds its own stamp to a user or server ID, to describe where in the organization they exist. This lets us know that ServerA is in Dallas, for example. The essential part of every ID file for every server and user in the same organization, however, will be the same CERT.ID. This allows authentication, which is discussed more in Chapter 3, "Notes Security".

Naming in Notes can be either flat or hierarchical. Flat names have a person or server known only by the given name—for example, Libby Schwarz or ServerA. No two names can be the same with flat names. Unfortunately, similarities in given names or functions can sometimes make it difficult to make unique names. Two people cannot have the name *JohnS*, for example, or two servers cannot be called *ServerA*. Furthermore, flat names are not too descriptive of the organization. A server with the name *Main* gives no information about the location or function of the machine. Although Notes continues to support flat naming for backward compatibility, hierarchical naming has been the default naming convention since Notes V3.

Hierarchical naming adds other levels to the given name (or common name) to make names unique. The extra levels are separated by a slash to show the hierarchy. For example, a hierarchical name might look like this: Libby Schwarz/Authors/TotalSem or ServerA/TotalSem. These extra levels help define the function or location of the name and allow redundancy in parts of the names while keeping the complete name unique. It is perfectly acceptable to have Libby Schwarz/Authors/TotalSem and Libby Schwarz/Instructors/TotalSem as part of the same organization.

NOTE
Lotus recommends hierarchical naming for all Notes installations. Notes 4.X has included some tools, such as the administration process (AdminP, described in Chapter 5, "Notes Replication"), to assist in automating the process of converting from flat names to hierarchical names. Most organizations, for a variety of reasons, are in the process of converting from flat names to hierarchical names.

One significant difference between flat and hierarchical names is the number of certificates contained in an ID file. In flat naming, each ID file has multiple certificates included within it. These certificates give the user access to servers. In hierarchical naming, each ID file has only one hierarchical certificate, although it may also have multiple flat certificates for communicating with servers in companies that use flat naming.

NOTE
A *certificate* is a stamp or code appended to an ID file in Notes by the process of registering (or certifying) the ID. The certificate contains an electronic signature that relates the ID to the certifier.

Hierarchical names use a tree structure to mimic the structure of your organization. Hierarchical naming is based on X.500 naming standards and is similar to the naming conventions used by Novell's NDS naming in NetWare v4. An example of a hierarchical name for a user includes the common name, one to four optional organizational units, and an organization. The organization, or O, will usually be your company's name. (Note that Lotus uses the term *organization* to describe both the whole naming scheme and the highest level in a hierarchical name.) Organizational units, or OUs, can be based on departments, geography, or other working units within your organization. The *common name*, or CN, will either be a user or server name. The other hierarchical unit that you will see used is a Country Code, which is based on the *International Standards Organization* (ISO) two-letter country codes. (These can be found in an appendix of the *Notes Administrator's Guide*.) The following paragraphs discuss each of the levels of the hierarchical name. Figures 1.2 and 1.3 show graphical examples of a hierarchical tree.

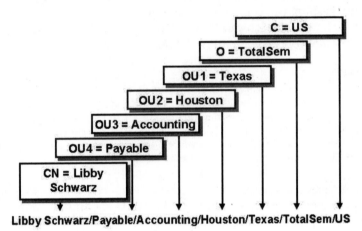

Libby Schwarz/Payable/Accounting/Houston/Texas/TotalSem/US

Figure 1.2   Hierarchical organization naming example

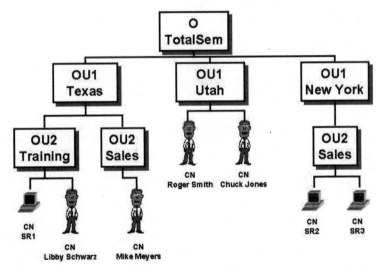

Figure 1.3   Hierarchical organization tree example

COUNTRY CODE

The *Country Code* (C)is an optional level of the hierarchical name. It contains a two-letter code specified by the ISO. The country code should be used only for organizations that have locations outside their country of origin. The country code, like all levels of the hierarchical name, will appear when a user's name is resolved, such as when sending e-mail.

ORGANIZATION

The *Organization* (O) is commonly the top level of a company's hierarchical naming scheme, as the country code is not used often. Frequently companies will use their company name—or an abbreviation of it—for the organization name, because it will be part of every name that is registered. The organization is created during the setup of the first server, when the CERT.ID file is created. This CERT.ID file, which will be stored by default on the first server, will be used to register either the organizational units or users and servers. This file should be secured and backed up, because it is one of the most important files in your Notes environment. With it, another administrator could create users and servers in your organization, if they had the necessary rights to the NAB. During the registration process, the organization is appended to each of the user and server's names, making it a required part of the hierarchical name.

Multiple organizations can exist in a single domain if necessary, although this option requires additional planning. Some reasons why this might be needed include a name change within a company or the need to connect to an external set of Notes servers. In a renaming situation where a new organization certifier was created, the administrator would need to recertify all the servers, users, and organizational units with the new organization certifier.

ORGANIZATIONAL UNITS

Organizational Units (OU) are optional levels in a hierarchical name that create more unique levels within your organization. OUs will appear beneath (or to the left of) the O name (or another OU) in the naming hierarchy.

Usually, OUs represent a division of your organization based on geography, department, or workgroup. There can be up to four vertical levels of OU, with any number of OUs at any given level. For example, if your first level of OU within your organization is based on geography, there can be any number of cities at that level.

Organizational units can either help ease the load on an administrator or increase it immensely. If the OUs are created without research into the patterns of movement in the company, every time a user transfers between departments, they might have to be re-certified. If OUs are used in a limited fashion, however, the users can make some movement without requiring re-certification. In addition, while four is the maximum number of vertical levels, most organizations will limit their use of OUs to one or two to ensure that they do not become an administrative burden. On the other side of the coin, OUs can assist an administrator in decentralizing what he or she has to do. In many organizations, and certainly in those that have only one certifier (the O), there would be only one or two administrators with access to this CERT.ID, and therefore, only one or two administrators would be available to register new users or servers. If multiple certifiers (OUs) are at lower levels, however, another administrator could take some of this burden. In our example company, Total Seminars, most of the administration is done in the central office in Houston. With geographically based OUs, however, an administrator in Dallas could be given the right to create users certified by the Dallas OU certifier, relieving some of the load on the central administrators.

In Notes hierarchical naming, there is no functional difference between the name of a server and the name of a user. There is no relationship required for servers and users to be registered under the same OU. All the servers in the Total Seminars organization might be certified by an OU called Servers, for example, that is directly under the O option. This would result in a server called ServerA/Servers/TotalSem. This would help to avoid the confusion of having a server called ServerA/Authors/Houston/TotalSem that would need to be used by everyone in the Houston and Dallas offices. Another option would be to have the server certified directly under the O or under the Houston OU. The biggest restriction is to be consistent and clear for the benefit of your users and administrators.

COMMON NAME

The *Common Name* (CN) is the actual name of the user or server. In the case of the user, you have fields for first name, middle initial, and last name. Only the last name field is required. In the case of a server, this is often the Domain Name Service host name or Net-BIOS name of the server, although this is simply for ease of administration. The common name can use up to 80 characters and would be the same name you would use as the complete name in a flat naming environment.

 NOTE
The DNS host name defines a server or workstation to the TCP/IP protocol and to TCP/IP applications. DNS host names require a DNS server, which resolves the name into an IP address. NetBIOS names define servers and workstations to Microsoft applications.

Table 1.1 gives a quick review of the vital information to remember about the various levels of the hierarchical name.

Table 1.1   Hierarchical Name Information

Element	Description
Country Code (C)	▪ Optional
	▪ 0 or 2 letters
	▪ Based on ISO standard 2-letter codes available from the Notes Administrator's Guide. Before using the country code, clear your company's O name with your country's clearinghouse for x.500 names.
Organization (O)	▪ Required for hierarchical naming
	▪ 3–64 letters or numbers
Organizational Unit (OU)	▪ Optional
	▪ Up to 4 vertical levels of OU allowed
	▪ 3 to 32 letters or numbers per OU
Common Name (CN)	▪ Required
	▪ Usually the person's first and last names and middle initial for a user or a server name for a server
	▪ 80 maximum letters or numbers

NOTE
For hierarchical naming, each user and server must have at least
a CN and an O.

Notes offers another hierarchical level, where two people end up
with the same hierarchical name. This is called the *User Unique
Organizational Unit* (UUOU). The UUOU adds a unique, descriptive
comment to a name to differentiate it from another name. There
are two engineers named Suzy Smith, for example, both in the
Austin office of Total Seminars. Their names would be identical:

```
Suzy Smith/Austin/TotalSem
```

A UUOU would allow the administrator to define them more in
some way, such as *Redhead Suzy* and *Brunette Suzy*. Now their names
would be unique in the organization:

```
Suzy Smith/Redhead/Austin/TotalSem
Suzy Smith/Brunette/Austin/TotalSem
```

To make a UUOU, select the Other panel of the Register Person
dialog box. One field enables you to add a UUOU, as shown in Figure 1.4. Note that a UUOU does *not* count as one of the four levels
of OU allowed, but it gives the administrator the capacity to create
some level of distinction between users with the same names, in
the same OUs.

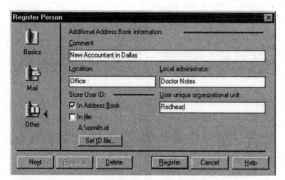

Figure 1.4 UUOU created while registering a user

Two more terms that you must know about hierarchical names
are fully distinguished abbreviated hierarchical names and fully dis-

tinguished canonicalized hierarchical names. Both contain exactly the same information but are written differently. A fully distinguished abbreviated name displays a name with the organizational levels separated by slashes. The levels are assumed by their place in the name. ServerA/Houston/TotalSem, for example, describes the server machine named ServerA, which is in the Houston organizational unit in the TotalSem organization. A canonicalized name spells all this out in the description. That same server written canonically, for example, would look as follows:

```
CN=ServerA/OU=Houston/O=TotalSem
```

Notes uses canonicalized names internally; however, humans generally use abbreviated names. The exam requires you to know both.

The advantages of hierarchical naming schemes clearly outweigh the minor disadvantages of requiring more setup time and thought, and possibly more administrative time to maintain. If done properly to begin with, a good hierarchical naming scheme can grow and evolve along with your enterprise. The advantages are spelled out in Table 1.2.

Table 1.2 Hierarchical Naming Scheme Advantages

Advantage	Description
Fewer duplicated names	In flat naming, you will often have multiple users with the same name. This requires the users to include numbers in their names or to use some other method to make the name unique. By appending organizational unit information, the users are more likely to be unique within the smaller unit.
Ability to decentralize administrative tasks	In flat naming, often one administrator does all the work of registering users and servers. When you use OUs, you have the ability to give certifiers, which are specific to business units or geographical locations, to the administrators at those locations. This will allow them to register the users and servers in those units.

continues

Table 1.2 Continued

Advantage	Description
Better security	When using hierarchical naming, you will have names that are more likely to be unique. This allows you to be very specific in Access Control Lists, Groups, and Server Access lists. In flat naming situations, you often will have users or servers with the same names. When this is the case, you may have users or servers that are accidentally given access to the wrong information.
Ability to use advanced Domino and Notes functionality, such as the Administration Process (AdminP)	AdminP and other advanced Domino services will recognize only hierarchical names.

Domains

A *domain* defines pathways for e-mail, the location of a central repository for addresses and configurations, and many other administrative functions. Domains are defined by the *Public Name and Address Book* (NAB), a database that holds much critical data, such as user and server names and configurations.

Domains function in Notes as containers for many servers, databases, and users, and the information is all interconnected. Typically, there should be only one domain in an organization for simplicity. Similarly, there is usually only one organization in a domain—again, for simplicity. Notes Release 4 can handle a massive number of users in one domain—up to 150,000—although there are other limiting factors that make the overall domain size smaller. A domain is not limited by geography or connections.

NOTE
A Notes domain is a completely separate issue from both Internet
and Windows NT Domains—although there are some relation-
ships and comparisons that can make administration a little easier.

The NAB holds all the critical information about the users and servers in your Notes Domain. The NAB contains configuration information about your Notes servers and the e-mail routing information needed to get e-mail from one user to another. The NAB is created when the first Notes server is installed in an organization. It is important to note that although the NAB is one database, there can be (and usually are) multiple replicas of the NAB scattered throughout all Lotus servers within the domain. The NAB contains many critical pieces of information, including:

- *Connection Documents,* which contain information used when connecting to other servers
- *License Documents,* which track the number and types of Notes licenses
- *Server Documents,* which define the configuration for all servers in the domain
- *Certifier Documents,* which track all certifiers in the domain
- *Person Documents,* which contain information on all users in the domain
- *Group Documents,* which create and define groups that can be used in mailing and security

One NAB per domain and one domain per NAB always exist. The primary purpose of a domain is to organize and define mail routing paths. We will describe the NAB more in Chapter 2, "The Notes Name and Address Book."

NOTE
It is important to understand the differences between a domain
and an organization in Notes. An organization's primary purpose
is naming and security. The primary purpose of a domain is e-mail
routing.

Notes Named Network (NNN)

A Notes domain is often spread among many far-flung geographical locations, employing many different types of network topologies.

Consider a company with two departments in different offices on the same floor of the same building. If one user sends e-mail to another user, the users are in the same domain and are probably on the same Ethernet network. So, the mail is sent to the other user. There is a lot of traffic going on between those departments. Now, take those two departments and separate them into different cities. The two networks are now connected by a modem. How can the modem connection be controlled so that mail is sent only at certain intervals? What if the two departments are on the same network but are running different network protocols? These problems require a special subdivision of the domain into groups that share constant connections and common protocols. These subgroups of the domain are called *Notes Named Networks* (NNN) and are an important feature for handling network connections that are *not* constantly connected and/or use different protocols. To see how an NNN provides this service, let's mentally subdivide the domain into two separate NNNs called DeptA and DeptB. When an e-mail is sent to a server, the mail server compares the NNN of the sender to the NNN of the recipient. Remember, all servers have a copy of the NAB. If the NNNs match, the message is sent to the recipient immediately. If they are different, the server checks for a connection document to tell it how to send the e-mail to the other NNN. This feature gives administrators control over the use of non-constant connections. Consider the two NNNs again. This time, they are all on the same hard-wired network, but the two servers use different protocols. There is another server that can handle both protocols. By checking the NNNs and cross-referencing the proper connection documents, the sending server knows to route the mail through the server that can handle both protocols. Clearly, NNNs are a powerful component for Notes mail routing.

Additionally, the NNN determines what servers a user sees in the File . . . Open Database dialog box, or what servers an administrator sees in the Server Administration panel. The servers in your NNN will be visible in those dialog boxes. To access other servers, you must type the distinguished name of the server.

NNNs are determined by two characteristics: protocol and constant connectivity. Servers that are in constant contact, such as same *Local Area Network* (LAN) or bridged/routed *Wide Area Network* (WAN), and share the same protocol can be in the same NNN. The two servers described in Figure 1.5, for example, are both on the same LAN, and both use the same protocol. They probably would be in the same NNN, because the servers fulfill both requirements.

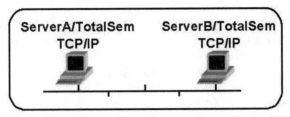

Same Protocol, Constant Connection - Same NNN

Figure 1.5 Same protocol, constant connection—same NNN

If the two servers were connected via modem, however, regardless of whether they both used TCP/IP, they would be in different NNNs because they are not constantly connected, as shown in Figure 1.6.

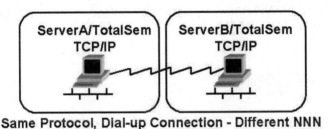

Same Protocol, Dial-up Connection - Different NNN

Figure 1.6 No constant connection—different NNNs

Similarly, two servers connected via a LAN, using different protocols (TCP/IP and IPX/SPX), would be in different NNNs because they do not meet the same protocol requirement. Servers that have multiple protocols will be in multiple NNNs (see Figure 1.7).

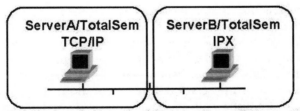

Different Protocol, Constant Connection - Different NNN

Figure 1.7 Different protocols—different NNNs

ServerA/TotalSem, for example, uses the TCP/IP protocol. ServerB/TotalSem uses the IPX/SPX protocol. They are connected to the same LAN. Because they use different protocols, they will not be in the same NNN. ServerC/TotalSem uses TCP/IP but is connected only when dialed in using a modem. It will be in a third NNN. When ServerD/TotalSem is installed, using the TCP/IP protocol and attached to the same LAN as ServerA/TotalSem, it will probably share the same NNN as ServerA/TotalSem.

NNNs are usually named for both their location and protocol, to avoid confusion. ServerA/TotalSem and ServerD/TotalSem are both on a LAN in Houston, using TCP/IP. An appropriate name for this NNN might be HOUSTON_IP or something similar.

To determine the NNN for a server, you can either specify this at the time when the server is created by changing the Network field from Network1 (the default) to a more descriptive name. The Network field is on the Advanced tab when configuring a new server, as shown in Figure 1.8.

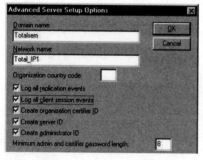

Figure 1.8 Assigning a NNN while configuring the server

Often an administrator does not know all the information necessary to determine the proper NNN when configuring the server, or the administrator may need to change the NNN after the server has already been installed. Another way to assign a server to the appropriate NNN is in the server document. The Network Configuration section of the server document, as shown in Figure 1.9, allows an administrator to specify the protocol and NNN that the server should use. Note that you can assign a server to use more than one protocol and NNN.

▶ Server Location Information
▼ Network Configuration

Port	Notes Network	Net Address	Enable
IP	Total_IP	Notes01	● ENA
IP	IP	Notes01	○ ENA
IP	IP	Notes01	○ ENA
IP	IP	Notes01	○ ENA

Figure 1.9 Assigning a NNN using the Server document

The main purpose for NNNs is mail routing. NNNs will be discussed again, therefore, in Chapter 4, "Notes Mail." NNNs also determine, however, which servers will be visible in File . . . Database . . . Open dialog lists and in the lists for the Server Administration panel. Administrators can use NNNs to display the appropriate servers to groups of users.

Installing and Configuring the First Server

The actual installation of a Notes first server is straightforward. The installation will install both the Notes server software as well as a copy of the Notes client software. The Notes client is required, because it is where the majority of the configuration settings will take place. After accessing the CD-ROM or other installation media, choose the correct folder on the CD-ROM for your environment. For installation on Windows NT on an Intel platform, choose the W32intel folder. Choose the Install subfolder and the INSTALL.EXE application to begin the Installation program. After accepting the licensing information, the installation program will begin.

 NOTE
The configuration screens vary in appearance from Notes R4 to R4.6, although the basic information being requested is the same. The following screen captures are taken from Notes R4.6. Note the use of the word *Domino* instead of *Notes Server*. Remember that the Lotus Exams still use the term *Notes Server*—so do not let that confuse you.

First, type your name and company, as shown in Figure 1.10. Click Next to confirm your entries.

The next dialog box enables you to choose whether to install a Notes server, Notes Mail Server, or to perform a Custom Install, as well as to choose the drives and folder locations for the software. Choose either Notes Server install or Customize features—Manual Install to install the complete server. By default, the software will install on the C:\NOTES and C:\NOTES\DATA directories, as shown in Figure 1.11. The executable files and dynamic link libraries, for example, are stored in the \NOTES directory. Files that are used to identify the server, templates, and other working files will be stored in or under the \NOTES\DATA directory. The NOTES.INI file will also be created on the machine. In a Windows NT installation, this file is created in the \WINNT subdirectory.

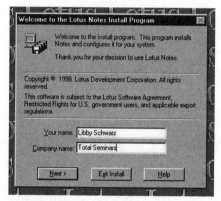

Figure 1.10 Installation screen

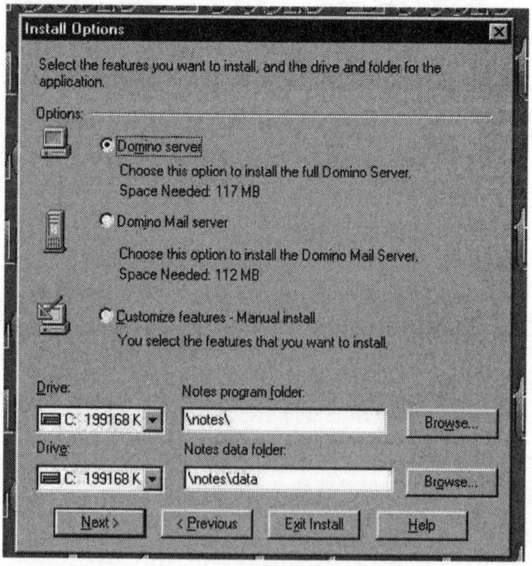

Figure 1.11 Installation options screen

If you choose a custom install, the Customize installation screen will be displayed to enable you to choose the specific options you want. This feature is shown in Figure 1.12. At a minimum, you must install the Workstation, Server, and data files. To install and use the advanced services, such as clustering, billing, and partitioned server available in R4.X, you must buy a separate license. Advanced services are not covered on either System Administration exam.

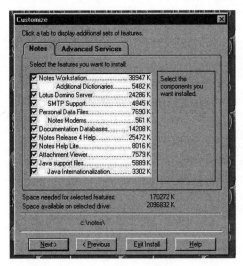

Figure 1.12   Customize installation screen

You will have the option to place the program folder and icons in an appropriate location, and then the installation program will begin copying files to the hard disk. In most environments, it is necessary to reboot the machine after the installation process and before beginning the configuration process. When the installation of the server is complete, it is time to start configuring the server.

To begin the configuration process for a server, you must first start the client (workstation) software on the server machine. This is true for all servers, not just the first one in the organization. When you initialize the client software on the server machine, you will be prompted to configure the server. The first question is whether this is a first server or an additional server in your organization. In this case, choose First Server, as shown in Figure 1.13.

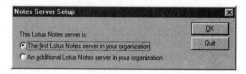

Figure 1.13 Setup screen

For the sake of comparison, Figure 1.14 shows the equivalent screen in Notes R4.6. The information is the same, even though the screen is quite different.

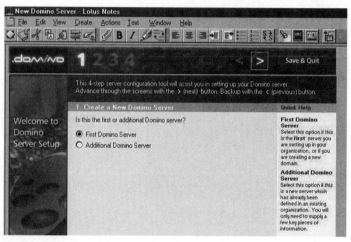

Figure 1.14 Installation screen in R4.6

The Setup screen, shown in Figure 1.15, enables you to configure the Organization, Domain, Server, and Administrator. When you type names for the Server, Organization, and Administrator on the first server, these elements define the organization and documents in the NAB. To define the domain, NNN, and country code (if necessary), choose the Advanced Options button.

By default, when the first server is configured, the certifier ID file is one of the files created automatically. This CERT.ID is the O certifier for your organization. In addition, a SERVER.ID for the first server is created, as well as a USER.ID for the administrator. On the Advanced Options screen, as shown in Figure 1.16, you have the option to prevent these files from being created. An administrator might choose to do this in a situation where the server was being rebuilt for some reason, but backup copies of the files already existed. In most cases, the organization certifier, server, and administrator ID files will be created during this part of a first server configuration.

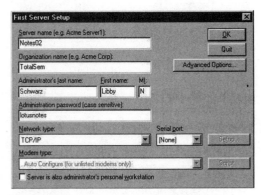

Figure 1.15   First server configuration screen

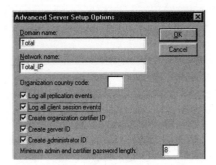

Figure 1.16   Advanced options configuration screen

 NOTE
To place the server or administrator IDs under a level or more of OUs, you would have to recertify the server and administrator ID files later, using OU certifier ID files. The steps for creating the OU certifier ID files are described later in this chapter.

In addition, when the first server in the domain is created, the NAB for the domain is created. The NAB corresponds directly to the domain—each domain has one NAB. The domain name is defined in Advanced Setup options. If you do not type in a specific domain name, Notes will use the same name for both your organization and your domain. This is common. After the options on these dialog boxes are completed, click OK to begin the process of registering and configuring the first server. It is during this period that the files described earlier will actually be created and placed in the \NOTES\DATA directory of the first server. The bar shown in Figure 1.17 indicates the progression.

Figure 1.17 Monitor the configuration

After the configuration process is complete, the server software can be initialized. When this occurs, additional files are created. Some of these files include the following:

■ ADMIN4.NSF. Used for the Administration Process discussed in Chapter 5, "Notes Replication."

■ LOG.NSF. Used to make a record of all actions taken on or by the server (discussed in depth in Chapter 7, "Advanced Configuration and Setup").

■ BUSYTIME.NSF. Used for searching free time with Calendar use.

Ports and Protocols

Additional decisions regarding ports, protocols, and Notes Named Networks need to be made almost immediately after starting a server. For the server to communicate with other servers and client machines, ports and protocols must be configured correctly. The Notes server uses the protocols used by the operating system. In our previous examples, TCP/IP is the protocol in use. Other common protocols include IPX/SPX and NetBIOS. Assuming the protocol is configured correctly in the operating system, you should choose the correct protocol while completing the first server setup, under the Network Type field.

Additionally, the Server document in the NAB has a section that describes the port, protocol, and NNN. The Server document is shown in Figure 1.18. In the Network Configuration section, the fields for Port and Network must be complete for the server to communicate. For example, for the protocol TCP/IP, the port will be TCP/IP.

▶ Server Location Information

▼ Network Configuration

Port	Notes Network	Net Address	Enabled
ᶠTCPIP↲	ᶠHOUSTON_IP↲	ᶠnotes02.totalsem.com↲	⦿ ENABLED ○ DISABLED
ᶠLAN0↲	ᶠHOUSTON_NB↲	ᶠNotes02↲	⦿ ENABLED ○ DISABLED
ᶠ↲	ᶠ↲	ᶠNotes02↲	○ ENABLED ⦿ DISABLED
ᶠ↲	ᶠ↲	ᶠNotes02↲	○ ENABLED ⦿ DISABLED
ᶠ↲	ᶠ↲	ᶠNotes02↲	○ ENABLED ⦿ DISABLED
ᶠ↲	ᶠ↲	ᶠNotes02↲	○ ENABLED ⦿ DISABLED

Figure 1.18 Network configuration section of the Server document

You can configure a port for a workstation by accessing the user preferences. Choose File . . . Tools . . . User Preferences from the menus. Select the Ports panel from this dialog box, as shown in Figure 1.19. This dialog box enables you to choose which ports should be enabled. If you are using the workstation software on the server machine, the server software must be down to enable or disable a port on that machine.

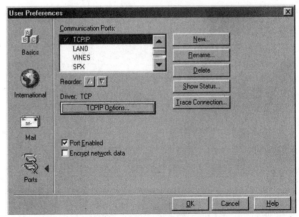

Figure 1.19 Enable the TCP/IP port using the User Preferences.

The Ports panel within the workstation software allows an administrator to do the following tasks:

- Enable and disable communication ports.

- Select the order in which the ports for a workstation or server should be used.

- Create a new port that uses a particular network driver, such as XPC for a modem using Com1.

- Choose to encrypt all the data sent over a particular network port (see Chapter 3, "Notes Security," for more information).

- Check the status of an enabled port to determine how it is functioning, as shown in Figure 1.20.

- Trace the connection to another server using a particular port. Tracing connections for troubleshooting is discussed in Chapter 6, "Administration Tools and Tasks."

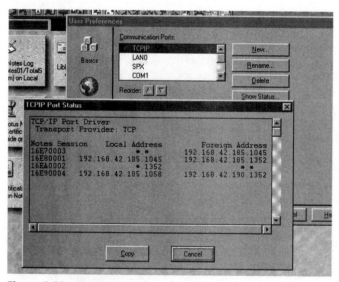

Figure 1.20 Show port status

> **NOTE**
> For more information on configuring and troubleshooting ports and protocols, please refer to the official Lotus documentation *Network Configuration Guide in* the Notes Administration Help database.

The Notes Network field refers to the NNN in which the server will reside. If this is the first Notes server, you can enter any name you wish. If this is an additional server, you can add the name of an existing NNN or create a new one. Remember that the default name for a new NNN is Network1, regardless of the protocol or connections. You must change the NNN for the server to communicate correctly.

Registering Organizational Units

A Notes server by itself is useless. We now need to begin the process of adding servers and users. Before we do this, however, we should first create Organizational Unit Certifiers that will allow us to register users and servers under the organizational units we created. Without them, you will be able to create a user like Mike Meyers/TotalSem, but you will not be able to make a Mike Meyers/Accounting/TotalSem profile. It is incorrect to use the term *creating* when referring to the process of making organizational units, servers, or users in Notes. The correct term is registering. Registering does more than simply add a server or user to a database—it also creates the ID files needed for authentication. To register organizational units, servers, and users, use the Files . . . Tools . . . Server Administration screen, called the Server Administration panel, as shown in Figure 1.21.

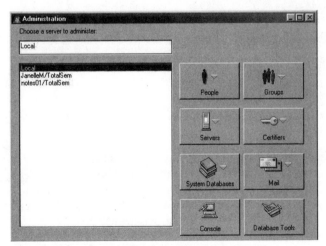

Figure 1.21 Administration main screen

To create organizational unit certifiers, click the Certifiers button and select Register Organizational Unit. The screen displayed in Figure 1.22 will appear.

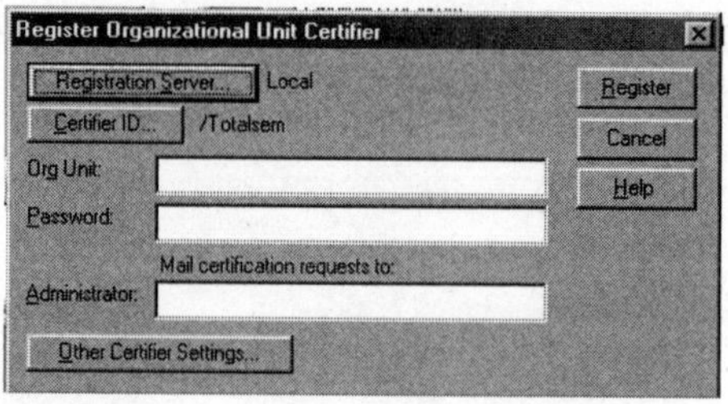

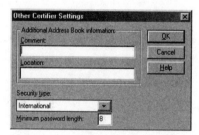

Figure 1.22   Certifiers main screen

Note that the name of the organization—TotalSem—is shown. Any organizational unit made will be a first-level OU under TotalSem. In other words, we can now make a Sales/TotalSem OU, but we cannot make a Houston/Sales/TotalSem yet. Let's register the Sales OU by simply typing the name Sales in the Org Unit field. You then enter a password for the OU and the name of the administrator for the OU. The Other Certifier Settings button, while optional, has some important features (see Figure 1.23).

Figure 1.23   Other Certifier Settings dialog box

The Other Certifier Settings enables you to enter a Comment and a Location if desired. You also choose the level of security here. Use *International* only if the installation will be outside North America. You can also set the minimum password length here. You can even set the password length to zero for no password, although this is not recommended.

Certifiers are some of the most important files in your Notes organization and should be tightly secured.

Once you choose the settings for the OU, Notes will prompt you for a floppy diskette. The new ID file for the OU will be copied to the floppy disk by default. Putting the ID on a floppy, as opposed to the hard drive, gives an extra level of security. Now that the OU has been certified, users and servers can be registered using that OU. To make a lower OU underneath the one you made, click on the Certifier ID . . . button again. It will prompt you for the proper certifier ID file. Insert the floppy with the newly created OU certifier ID file rather than the O certifier, enter the password for the OU, and the OU/O name will appear. You can then register another level of OU underneath the one you have already created.

Registering Users

Registering users is similar to the process of registering OUs. From the Administration screen, select the People button and choose Register Person. You will be prompted for a password for the current O or OU, after which Notes displays the Register Person dialog box, shown in Figure 1.24.

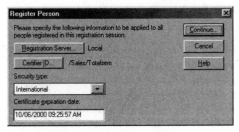

Figure 1.24   Register Person dialog box

On this screen, you select the server that will register the user and store the user's information, as well as the OU to which the person belongs. To choose a different OU or to use the O to register a user, click the Certifier ID button. Choose the O or OU certifier ID that you want to use for the registration process. Use the same process shown earlier to register lower OUs to make the user a member of a particular OU. This screen also enables you to choose the proper security type for the ID that you are creating, either International or North American. Again, choose North American for tighter security, unless the organization will be used outside the continent. You should use a single security type throughout your organization for best results. You would then click the Continue button to see the next Register Person dialog box, as shown in Figure 1.25.

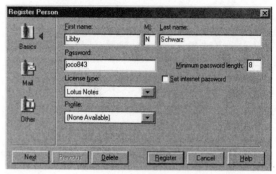

Figure 1.25 Continue using the Register Person dialog box

Three panels are available in this dialog box: Basics, Mail, and Other. The Basics panel, shown in Figure 1.25, enables you to enter the user's name, default password, and type of license. The profile is a document that holds some of the information necessary to configure a new user. This feature is optional, but in some cases it is extremely handy (refer to Chapter 6, "Administration Tools and Tasks").

> **NOTE**
> A user that is given a Notes license can create and use all the elements of Lotus Notes, including all types of design and administration. A user that is given a Notes desktop license can create and use all types of databases in Notes but cannot do any design or administration work. A user that is given a Notes Mail license can use only databases, such as their mail, NAB, and journal, that are registered for use by mail users. This is a limited license. Before creating users, determine which licenses your organization has purchased and what types of needs the users have. For more information on the databases available for users with a Notes Mail license, refer to the Notes Help database.

The Mail panel (Figure 1.26) defines the type of mail—Lotus Notes by default, the name of the user's mail file, and the name of the server that will hold that file. You can have the mail file created immediately or when the user sets up the Notes workstation.

The last panel, Other, is important. You can add Comments and Location information in this panel, but the important option is the Store User ID. The User ID contains all of the user's certificate information, as well as the information necessary to allow the user to authenticate with servers in your organization. Without this file, the user cannot configure the Notes workstation or use your Notes environment. The user ID can be attached to the Person document in the NAB and/or placed in a file called USER.ID on a separate disk or drive. The most common method is to place the user ID on a separate disk or drive. You will then give that file to the user or to the person setting up the Notes workstation. This file will then be used during the installation of the workstation, to allow the user to authenticate with the server. If the User ID is left in the address book, anyone who knows the username and password can access that person's ID file and other information.

Figure 1.26   Mail panel of the Register Person dialog box

> **NOTE**
> A good administrator creates new users with a default password and keeps a copy of every USER.ID file in a safe place. The users will then change their passwords when they configure their workstations. But if a user loses the password, the administrator can give the user the backup copy of the USER.ID file with the default password. Without it, the user must be recreated, and some data may be lost.

The other interesting option in the Other panel is the UUOU, mentioned earlier in this chapter. The UUOU allows the creation of a special OU for those rare occasions where the regular OUs are not enough to make a unique name (two people with the same name in the same OU). Figure 1.27 shows the Other panel.

Most Notes installations need to register hundreds or thousands of users. Imagine the magnitude of registering all users in the domain using the previous process. Fortunately, Lotus provides another method for registering users using text files. Most organizations have databases of some sort that contain the names and other information for their users. By database, I am not necessarily talking only about databases in database programs such as Access or

Paradox, nor of databases such as a Notes database. In this case, a database could be a listing of users in an Excel spreadsheet, ACT! Contact list, or WordPerfect Mail Merge file. The one thing that all of these programs have in common is the capacity to output their data in a *delimited* text format. Delimited means that all of the data is separated by a special character, such as a comma or a semi-colon, called a delimiter. Notes prefers to use the semi-colon for a delimiter, but commas and tabs are also commonly used. To use a text file to register a group of users, the data must be in the following format, with each line of the text file being a different record (user):

```
Lastname;Firstname;MiddleInitial;organizational
unit;password;IDfiledirectory;IDfilename;homeservername;
mailfiledirectory;mailfilename;location;comment;forwarding
address;profile name;local administrator
```

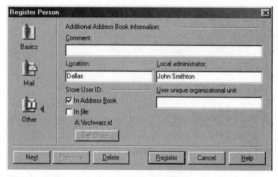

Figure 1.27 Other panel of the Register User dialog box

 NOTE
Sometimes the information will wrap to a second or third line, as shown here. The records for users are determined based only on the new lines created by pressing Enter, not by wrapping.

The fields used in the registration batch file are explained in the following paragraphs:

- Last Name—required. Used to specify a last name.

- First Name, Middle Initial—optional. The user's first name and middle initial.

- Organizational Unit—The OU to which the user will be assigned.

- Password—The user's default password.
- ID File Directory—Location of the user's ID file.
- ID File Name—Name of the user's ID file. If a name is not provided, Notes will create an ID file based on the user's name.
- Home Server Name—Home server for the user's mail.
- Mail File Directory—Directory for the mail file on the home server.
- Mail File Name—Name of the user's mail file. If a name is not provided, Notes will create a mail file based on the user's name.
- Location, Comment—The Location and Comment fields for the user.
- Forwarding Address—Used for other Internet mail only.
- Profile Name—Specifies a user setup profile.
- Local Administrator—The user's local administrator.

Only the last name and password fields are required. If you choose to skip any field, you must still add the delimiter for that field. All records in the delimited text file must hold the same information. Here are three examples taken from Lotus Notes Help.

Example 1: Text file specifying last names and passwords only.

```
Abbot;;;;password1
Rosen;;;;password2
Smith;;;;password3
VanDorn;;;;password4
```

Example 2: Text file specifying complete names, passwords, home servers, and user setup profiles.

```
Abbot;John;A.;;password1;;;MARKETING/ACME;;;;;;Marketing
Profile
Rosen;Alex;S.;;password2;;;SALES/ACME;;;;;;Sales Profile
Smith;Erica;A.;;password3;;;MARKETING/ACME;;;;;;Marketing
Profile
VanDorn;Betty;;;password4;;;SALES/ACME;;;;;;Sales Profile
```

Example 3: Text file specifying more information for each user.

```
Abbot;John;A.;;password1;C:\USERID;JAABBOT.ID;MARKETING/ACME;MA
IL;JABBOT;Marketing;#53551;;Marketing Profile;Susan McKay
Rosen;Alex;S.;;password2;C:\USERID;ASROSEN.ID;SALES/ACME;MAIL;A
SROSEN;Sales;#47809;;Sales Profile;Jan Wong
```

```
Smith;Erica;A.;;password3;C:\USERID;EASMITH.ID;MARKETING/ACME;M
AIL;ESMITH;Marketing;#80133;;Marketing Profile;Susan McKay
VanDorn;Betty;;;password4;C:\USERID;BVANDORN.ID;SALES/ACME;MAIL
;BVANDORN;Sales;#56778;;Sales Profile; Jan Wong
```

Once the text file has been created, users can be registered into your Notes domain and organization by opening the Server administration panel and choosing People . . . Register from File. Notes prompts for a password and certifier and displays the screen shown in Figure 1.28. This is the same screen you complete when registering users individually. After you select continue, Notes prompts for the text file name and then registers the new users.

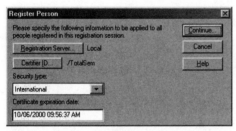

Figure 1.28   Register Person dialog box

Registering Additional Servers

The process of registering additional servers is similar to registering users. After you open the Server administration panel and choose Servers—Register server, you see the same dialog box displayed in Figure 1.26. Choose the registration server and the proper certifier. Notes then displays the Register servers dialog box. On the Basics panel, shown in Figure 1.29, complete the server name, default password, minimum password length, domain, and administrator. In most cases, you will type in the name of the existing Notes domain in the domain field. Notes requires a password of a minimum length of one character if you are going to store the server ID in the NAB.

On the Other panel of the dialog box, shown in Figure 1.30, the most important items are the Network field and the location to store the server ID file. The default Network name is Network1, as we mentioned earlier. Change the NNN name here to ensure proper communication within your Notes domain. Similarly to when you register a user, Notes creates an ID file that is used when installing the new server.

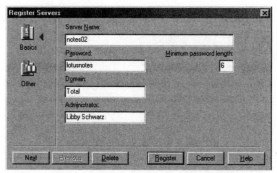

Figure 1.29 Basics panel of the Register Servers dialog box

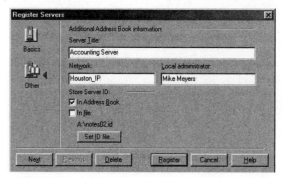

Figure 1.30 Other panel of the Register Servers dialog box

Installing Additional Servers

The few steps of installing an additional server are identical to installing the first server. Select the proper directory, the proper INSTALL.EXE file, and the type of installation desired. The big changes come after the installation, when you configure the server by running the workstation software for the first time. Instead of selecting the First Lotus Notes server . . . option, select the Additional Lotus Notes Server . . . option. The dialog box shown in Figure 1.31 appears.

On this screen, you type the new server's name, the name of the server from which it should obtain a replica of the NAB, and the protocol that the new server will use. Here, you also can select to obtain the server's ID from a file, if supplied. If you do not supply an ID file for the server, it will look for one in the server document in the NAB.

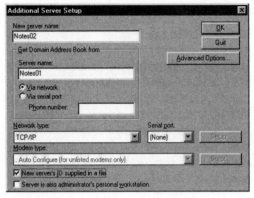

Figure 1.31 Configure an additional Notes server.

The Advanced Options button has three choices:

1. Log All Replication Events, which will write any replication events that take place both to the server console and to the Notes Log (LOG.NSF).

2. Log All Client Session Events, used to track each user's access to the server on the console and in the Notes Log.

3. Administrator's ID is supplied in a file, to add the Administrator if a file is provided.

Older versions of Notes (4.X–4.5) do not have the last option and instead will have a Log Modem I/O option for logging any responses from an installed modem.

Breaking Down a Server

An important trick for a Notes System administrator is the function of *breaking down* a Notes Server. Breaking down a Notes server is the process of removing the files that define the server in the organization and in the domain, as well as the files that are used to configure the server. After you break down a server by removing these files, you can run the configuration program again without reinstalling the software.

You might need to break down a server in your Notes environment for the following reasons:

- If you need to change the server's name
- If you need to recertify the server into a different O or OU
- If you need to create a new domain in your environment
- If you make a mistake when configuring a server the first time

To break down a Notes server, you will remove the following files by deleting them. You might want to move them to another folder or drive, in case you realize later that you need some information from them. The files that need to be deleted are all stored in the \NOTES\DATA directory by default:

- DESKTOP.DSK
- CACHE.DSK
- NAMES.NSF**
- MAIL.BOX
- CERTLOG.NSF
- LOG.NSF
- CATALOG.NSF
- ADMIN4.NSF
- STATREP.NSF
- EVENTS4.NSF
- BUSYTIME.NSF (Notes 4.5 and higher)
- \MAIL subdirectory and all its contents
- CERT.ID file**
- SERVER.ID file*
- USER.ID file (the administrator's user ID)*
- MAILOBJ.NSF and the shared mail databases (if shared mail is enabled)
- JOBSCHED.NJF

You may not want to delete the files that are marked with a single asterisk if you are not planning to reregister the user or server. You may not want to delete the files that are marked with two asterisks if you are not recreating the domain.

NOTE
Do *not* delete files that have an .NTF extension. If you are not viewing the extensions for all file types, you will probably see two identical files with the same names for most of the files listed earlier. One of these files is a database (.NSF), and the other one is a template (.NTF). If you delete the template files, Notes will be unable to create the necessary databases during the configuration process, and you will have to reinstall the template files from the original media.

In addition, you will need to edit the NOTES.INI file. This file is usually stored in the \WINNT subdirectory in an NT Server installation, or in the \WINDOWS subdirectory in installations of other versions of Windows. Remove all lines from the NOTES.INI file except for the following lines:

```
[NOTES]
DIRECTORY=C:\NOTES\DATA
KitType=2 (servers) or 1 (workstations)
```

 NOTE
The process for breaking down a Notes workstation is essentially the same as that for a server. Not all files listed previously, however, exist on a Notes workstation. Delete the ones that do.

Installing and Configuring a Notes Workstation

The process of installing and configuring a Notes workstation is similar to that of installing and configuring a server. Before you can install and configure a workstation, you must first have used the processes described previously to register the user that will use that workstation. In Notes, a workstation is configured for a particular user.

Before you begin the process, ensure that you have the user's ID file and that the user's mail file and person document are created and in the correct locations.

The installation process is the same as for installing servers. After accessing the CD-ROM or other installation media, choose the correct folder on the CD-ROM for your environment. For a client installation on Windows NT on an Intel platform, choose the Client . . . W32Iintel folder. Choose the Install folder and the INSTALL.EXE application to begin the installation program. After you accept the licensing information, the installation program begins.

In most cases, you will select either a Standard Install or a Manual Install to customize features, as shown in Figure 1.32.

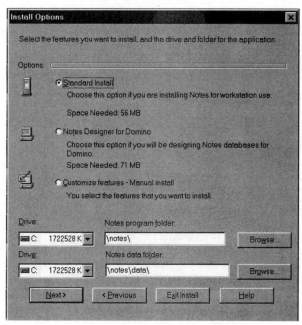

Figure 1.32   Choosing a client installation type

> **NOTE**
> If you are using the server software installation media (down-loaded from the World Wide Web, for example), use a Custom install and only choose to install the Notes client software and the Personal Data files.

If you select a Manual install, you will see the screen in Figure 1.33—which enables you to choose the elements to install. At a minimum, you must install the Notes Workstation and the personal data files. You may also select to install the help and documentation databases, the Notes modem files, and others.

Allow the installation program to copy the files to your hard disk. By default, the executable files and dynamic link library files are stored in the C:\NOTES directory. The databases and other configuration files are stored in the C:\NOTES\DATA directory. The NOTES.INI file is stored in your \WIN or \WINNT subdirectory. After the installation is complete, you can begin the configuration. In some cases, you may need to restart your computer before running the configuration.

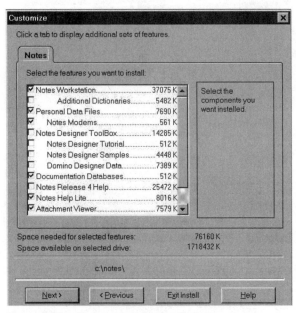

Figure 1.33 Customizing the installation

To configure the workstation, open the Notes workstation software. The first Notes Workstation Setup dialog box, shown in Figure 1.34, requires you to indicate the type of connection you have to your Notes server. This can be a LAN direct connection, a remote connection via modem, both, or neither. If your Notes user ID is in a file on a diskette, choose the Notes User ID has been supplied in a file option. You will be prompted to locate and select a USER.ID file. If you are obtaining your Notes user ID from the NAB, leave this option empty.

If your Notes ID is supplied in a file, you will be prompted to decrypt it by typing the password. The second screen in the Notes workstation setup process is shown in Figure 1.35. Your user name is already completed if your Notes ID was supplied. If you are obtaining your Notes ID from the NAB, you must type in your hierarchical user name in the User Name field, as well as the default password for the ID file.

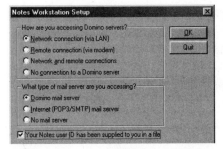

Figure 1.34   Notes Workstation Setup dialog box

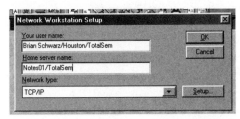

Figure 1.35   Continuing the Notes workstation setup

You will also type in the name of your home server. Remember that the user and server names should be hierarchical. Finally, select the one network type (protocol or port) to use to connect to the server. Before you select a network type, ensure that you know which protocol your Operating System uses and which protocol the server uses. You must use the same protocol that your server is using in order to communicate. After you click OK, the workstation setup process will create a personal NAB on your workstation as well as on other databases, depending on the version of Notes and the options in your User setup profile. In addition, icons for your mail file and the NAB are placed on your workspace. Finally, you must choose your time zone to complete the setup.

After you complete your workstation setup, you may want to add your preferences. To change any preferences on the workstation, select File . . . Tools . . . User Preferences from the menus. You may use this dialog box, shown in Figure 1.36, to enable or personalize a variety of options.

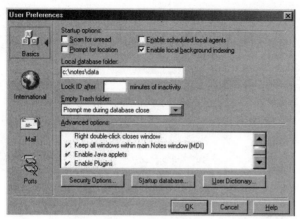

Figure 1.36   User preferences

User Preferences

Some of these options are listed here:

- On the Basics panel, you can select the default folder for your local databases, choose to lock your ID file after a certain time of inactivity, select when to empty the trash folder for your mail database, and select other options that determine the look, feel, and use of your workspace.

- On the International panel, you can choose whether to use imperial or metric measurements, the day of the week on which your calendar should start, and a dictionary for your language of choice.

- On the Mail panel, you can select some of the defaults for how your mail database will function. You can choose how often your workstation will poll for a new mail message, how your mail will be treated when you send it, and which address book to use for name resolution.

- On the Port panel, as described earlier in this chapter, you can select which ports Notes will use to communicate with your server. Choose the port that corresponds to your active protocol. You can choose more than one port to be enabled, and you can create new ports or delete existing ports. You also can reorder the ports. Notes uses the order listing to determine which protocol and port to communicate with first.

Review Questions

1. When planning her Notes environment, Tina wants to place her servers in NNNs. Where can she specify the NNN for a server?

   a. The Network document

   b. The Register Servers dialog box

   c. The Server document

   d. The Configure server dialog box

2. After deciding where to go to define the NNNs for her Notes environment, Tina asks for help in deciding how many NNNs her environment needs. She has four servers. Two of those servers are in Houston, connected via a LAN, using TCP/IP. One of the other servers is in Dallas, connected to the Houston network via a WAN, using TCP/IP. The last server is located in Kansas City, connected to the Houston network as necessary using a dial-up connection. It also uses TCP/IP. How many NNNs do you recommend?

   a. 1

   b. 2

   c. 3

   d. 4

3. What is the default name for an NNN, if you forget to change it?

4. What is the main purpose of an NNN?

   a. Mail routing

   b. Security

   c. Network connections

   d. Database access

5. What is the main purpose of a Notes domain?

   a. Mail routing

   b. Security

   c. Network connections

   d. Database access

6. What is the main purpose of a Notes organization?

   a. Mail routing
   b. Security
   c. Network connections
   d. Database access

7. Which of these files is created during the first server setup and configuration?

   a. NAMES.NSF
   b. LOG.NSF
   c. CERT.ID
   d. SERVER.ID

8. What should Harold use to register new users in his organization?

   a. CERT.ID
   b. An OU certifier
   c. SERVER.ID
   d. The NAB

9. Hierarchical names must have which components of the name?

   a. O and CN
   b. OU and CN
   c. C and O
   d. CN and OU1

10. How many hierarchical certificates will a server or user ID file contain?

    a. As many as they are given
    b. As many as there are levels of OU, plus one for the O
    c. One
    d. None. They are stored in the NAB, not in the ID file.

11. Which of the following would be the fully distinguished canonical name for a user whose common name is Joe Smith, and who was registered with the following certifier: Accounting/Houston/MLBConsult?

    a. CN=Joe Smith/OU=Accounting/OU=Houston/O=MLBConsult

    b. CN=Joe Smith/OU1=Accounting/OU2=Houston/O=MLBConsult

    c. CN=Joe Smith/OU2=Accounting/OU1=Houston/O=MLBConsult

    d. Joe Smith/Accounting/Houston/MLBConsult

12. What is the minimum number of characters an O can be?

    a. 1

    b. 2

    c. 3

    d. 4

13. Where do you enable a port for a server?

    a. File . . . Tools . . . Server Administration . . . Ports

    b. Server document . . . Port Configuration

    c. Server document . . . Network Configuration

    d. File . . . Tools . . . User Preferences . . . Ports

14. If a user needs to be able to create personal views, can you give them a Notes Mail License?

    a. Yes

    b. No

15. If a user needs to be able to design views, can you give them a Notes Desktop License?

    a. Yes

    b. No

16. What do you use to create a UUOU for a user?

   a. CERT.ID
   b. Register Person dialog box
   c. Person document
   d. An OU certifier

17. How many levels of OU certifier can you create?

   a. 1
   b. 2
   c. 3
   d. 4

18. If you want to register multiple users, what can you do?

   a. Use a Lotus 1-2-3 file and click Import on the file menu.
   b. Choose Next after each user is registered until you have registered all the users.
   c. Choose to register users from a text file.
   d. Create person documents by copying and pasting, then change the names.

19. If you make a mistake when configuring a Notes server, what can you do?

   a. Reinstall the software from the CD and run the configuration process again.
   b. Break down the server by deleting the SERVER.ID and NAMES.NSF files only.
   c. Break down the server by deleting the appropriate files and removing most of the lines in the NOTES.INI file.
   d. Break down the server by deleting the NOTES.INI file and NAMES.NSF.

20. When configuring a server, you should open which software first: Workstation or Server?

Answers

1. b or c. The NNN can be specified when registering a new server or in the Network Configuration section of the Server document.

2. b. The three servers that are constantly connected and use the same protocol will be in one NNN. The server that is not constantly connected will be in a second NNN.

3. Network 1

4. a. Mail Routing

5. a. Mail Routing

6. b. Security

7. a, c, and d. The LOG.NSF is not created until the server software is run for the first time.

8. a or b. Depending on the organization, you can either use the O, contained in the CERT.ID file, or an OU certifier to register users, servers, or other certifiers.

9. a. A hierarchical name requires an O and a CN component. All other components are optional.

10. c. The ID file will store the original hierarchical certificate. Any additional certificates are stored in the NAB. All flat certificates are stored in the ID file.

11. c. A canonical name requires all the elements to be included. It also requires OUs to be numbered.

12. c. Because a C is two characters, O certifiers are required to be a minimum of three, with a maximum of 64.

13. c. Server document . . . Network Configuration. Use File . . . Tools . . . User Preferences . . . Ports to enable ports for a workstation.

14. b. No

15. b. No

16. b. Register Person dialog box

17. d. 4

18. b or c will let you register multiple users, alathough c will save you lots of time.

19. a or c are equally correct, although c will save you more time.

20. Workstation

CHAPTER 2

The Notes Name and Address Book

As we described in Chapter 1, "Installation and Configuration," one of the elements of a Notes environment is the Notes Domain. The Notes Domain is all of the users, servers, and groups that share a single, common Notes Public NAB. This chapter will take you through many of the views and documents in the NAB, describing the purpose and use of each. In addition, each user has a Personal NAB that adds configuration information to their workstation. This chapter will also describe the views, documents, and functions of the Personal NAB.

Objectives

After reading this chapter, you should be able to answer questions based on the following objectives:

- Describe the purpose of the Notes Public Name and Address Book.

- Know the difference between the Public NAB and a user's Personal NAB.

- Describe the documents stored in the Public NAB.

- Describe the documents stored in the Personal NAB.

- Use documents in the NAB to create and control servers, users, groups, configurations, and connections.

The Notes Public Name and Address Book

The Public NAB is the most important database in your Notes environment and serves two main functions:

1. The NAB defines your Notes Domain and provides a directory of Notes servers, users, groups, and the organizational structure of your environment.

2. The NAB contains the documents that manage your servers and the connections and configurations.

These two functions mean that even the environments that do not use Notes Mail require a NAB for configuration and connections.

Most Notes environments use one NAB (and one domain) to contain their entire organization. This single NAB, which is created when the first server in the environment is configured, should be replicated to all other servers in your domain. This action ensures that any changes to users, servers, and configurations are shared throughout the environment. Although it is possible to have multiple domains and address books within a single company, it is recommended to use a single domain for an organization when possible. (The use of multiple domains is discussed in Chapter 7, "Advanced Configuration and Setup.")

As an administrator, one of your most important tasks is managing the Public Address Book for your domain. You must ensure that changes to the Public Address Book are made only by the correct administrators—and that those changes are replicated to each server in the domain.

As mentioned earlier, when you configure the first server in your domain, Notes creates a Public NAB with the file name NAMES.NSF. When you register and configure other servers, Notes places a replica of this NAB on those servers. You must then create the appropriate connection documents to ensure that changes are replicated correctly around your domain.

The NAB stores the following types of documents:

- Group documents
- Location documents

- Person documents
- Certifier documents
- Server Configuration documents
- Server Connection documents
- Domain documents
- Mail-in database documents
- Server program documents
- Server documents
- User Setup Profile documents

These documents are arranged into a variety of views in the NAB, which give you the information necessary to administer your domain.

NAB Documents

The following sections describe most of the documents that are stored in the Public NAB. The discussion includes the following concepts:

- The function of the documents
- The fields on the documents
- When and how the documents should be created

Some types of documents, such as Location documents, User Setup Profile documents, and Server Connection documents, are touched lightly here—as they are discussed in detail in other chapters.

 NOTE
The administration section of each of these documents contains fields for Owners and Administrators. These fields give editorial rights to these documents. Place users with Author access to the NAB in these fields, to enable them to edit these documents.

Group Documents

Group documents are lists that organize users, servers, and other groups into collections that have something in common. Groups can be used for mailing lists and for security. Two groups are automatically placed in the NAB when it is created: LocalDomainServers and OtherDomainServers. These two groups should contain the hierarchical names of the servers with which your server communicates (both within and outside your domain). To create a new group in the NAB, go to the Groups view and click the Add Group action button. The Group form, shown in Figure 2.1, is displayed. You can also choose Create . . . Group from the menus.

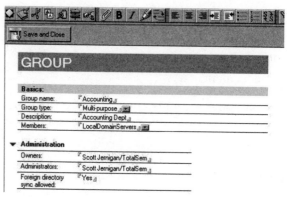

Figure 2.1 Group document

Give the group a unique name. The group name can have up to 64 characters and should not use spaces (for easier administration and mailing purposes). Your choices in the Group Type field include the following:

- *Multi-purpose.* These groups can be used for security and mailing. The names of these groups will be visible in the Address dialog when addressing mail and when adding users and groups to an access control list.

- *Access Control List only.* These groups can be used only for security. They are displayed in the Address dialog when you are adding users and groups to an access control list. They will not, however, be visible when addressing mail.

- *Mail only.* These groups can be used only for mail distribution lists. They are displayed in the Address dialog when you are addressing mail, but they are not available when you are adding users and groups to an access control list.

- *Deny List only.* These groups are visible only in the Deny Access Groups of the NAB and can be only used to deny access to servers.

Notice that when you change the group type field, the document title changes as well. You can place a description of the group's purpose in the description field. This field appears in the Groups view of the NAB and may help users to use the group correctly. Use the drop-down arrow to add users, servers, and other groups to the group Members field. If you choose to type a name instead of selecting it from the Address dialog, be sure only to use hierarchical names.

If a group becomes too large, you will not be able to save it after adding new members. If this problem occurs, you can create additional groups and nest these under the main group. If you create a group that contains all the users in your organization, for example, you might have to create a set of groups that contains employees from A-L and M-Z. These groups then would be placed inside the original group. You can nest up to six levels of groups.

Use either the Groups view or the Deny Access Groups view to see information contained in Group documents.

Location Documents

Location documents provide information for mobile users about their current environment. Location documents for each user are stored in the Personal NAB. The Location documents stored in the Public NAB are usually created by the administrator, so that their users can copy them into their Personal NAB instead of having to create them. We will discuss the fields on Location documents later in this chapter. View Location document information in the Locations view.

Person documents

Person documents describe and identify Notes users. The Person document contains information about the user's mail file and home server, as well as the public key and license information. When you register a new user into the environment using the Server administration panel, Notes automatically creates a new Person document for that user. You can also create a Person document manually by opening the People view of the NAB and selecting Add Person. You may also edit an existing user by choosing the Edit Person action button. A Person document is displayed, as shown in Figure 2.2. Use the following views to access Person documents:

- People
- Licenses
- Mail users

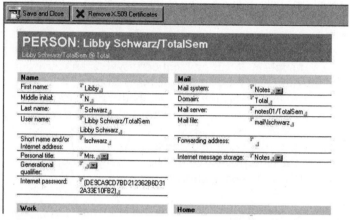

Figure 2.2   Person document

 NOTE
Just adding a Person document to the NAB does not do all the necessary tasks to register the person as a user in the domain. Users need to have mail files, certified user ID files, and other information created for them to be fully registered users in the domain.

The Name section of the Person document contains fields for the first name, middle initial, and last name. There is also a field for a Notes User Name. The Notes User Name field contains the official, fully distinguished name for the user that will be used in mailing and for security. Other entries in the User Name field are aliases for this user. Although you can add to or change the aliases if necessary when you edit a person document, the other fields should not be changed directly, because they are created during registration. To change a user's name, use the Rename Person action as described in Chapter 6, "Administration Tools and Tasks." There are also fields for an abbreviated short name, any generational qualifiers such as Jr. or Sr., and any personal titles such as Ms. or Dr.

The Mail section of the Person document defines the mail system that the person uses in the Mail System field. This field includes options for Notes mail, cc:Mail, Internet Mail (SMTP), X.400 mail, Other, and None. The default is Notes mail, but if you have the appropriate Notes add-in software, you can install these options and enable users to use other types of mail systems. This section also defines the user's mail domain and mail server. The name of the user's mail file is also listed in this section. Notes uses these fields to deliver mail to the user. If these fields are incorrect, the user will not receive mail.

The Work, Home, and Company information sections are not used for internal Notes purposes and are informational only. They can be completed or not, depending on the information your users would like to see in the NAB.

In the Misc section, the most important field is the Encrypt incoming mail field. The Notes router uses this field to determine whether it should encrypt the mail it is delivering to the user. Encrypting incoming mail will prevent any user except the recipient from reading the mail message. We discuss encryption more fully in Chapter 3, "Notes Security."

The Public Keys section contains a hexadecimal number that is the user's unique certified public key. This key is used for encryption, in combination with the private key stored in the ID file. The public key is also used with signing and authentication.

The Administration section contains fields for Owners and Administrators. These fields give editorial rights to these documents. Place users with Author access to the NAB in these fields to enable them to edit this document. Also contained in the Administration section is the Notes client license field. You can see and change the type of license being used by a particular user in this field. See Chapter 1, "Installation and Configuration," for more information on Notes licenses. Other fields in this section include the User Setup Profile (used with this user) and password checking configuration fields (r4.5 and higher).

Certifier Documents

Certifier documents contain information about certifiers in the organization. Certifier documents are created automatically when an O or an OU certifier ID is registered using the Server Administration panel. Certifier documents are also created during cross-certification, discussed in Chapter 7, "Advanced Configuration and Setup." Information about Certifier documents can be seen in the Certificates view, shown in Figure 2.3. The Certifier documents will contain the user's name, public key, the name of the issuer, or parent, of the certifier, and information about owners or administrators.

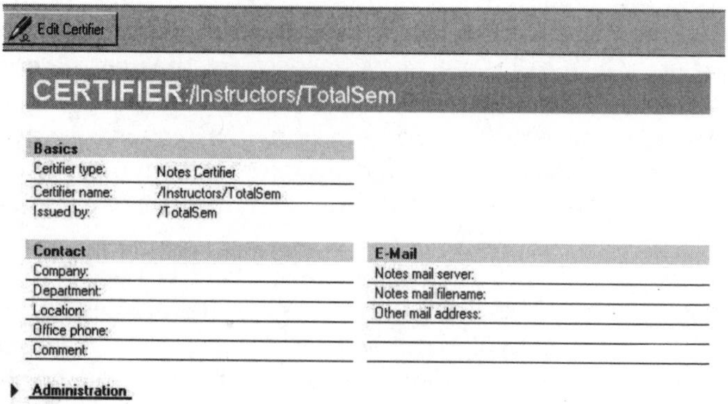

Edit Certifier

CERTIFIER:/Instructors/TotalSem

Basics

Certifier type:	Notes Certifier
Certifier name:	/Instructors/TotalSem
Issued by:	/TotalSem

Contact		**E-Mail**	
Company:		Notes mail server:	
Department:		Notes mail filename:	
Location:		Other mail address:	
Office phone:			
Comment:			

▸ **Administration**

Figure 2.3   Certificates view

Configuration Documents

Server configuration documents can be used to change or insert NOTES.INI settings. Creating a server configuration document is a safer way to work with the settings in the NOTES.INI file than opening the INI file and editing it directly.

NOTE

Not all settings available in the NOTES.INI file can be edited or changed using server configuration documents. Additionally, sometimes server configuration documents do not make the configuration changes to the server immediately—they may require the server to be rebooted first.

To create a Configuration document, open the Configurations view and click the Add Configuration action button. You can also edit an existing server configuration document if the configuration settings all apply to the same server or servers. A server configuration document is shown in Figure 2.4.

The server name field should be completed with the hierarchical name of the server or servers to which the configuration setting will be applied. You can place an asterisk in the field to indicate that the configuration setting applies to all servers in the domain. (In R4.6, you create a separate document for domain-wide configuration settings.) You may also list a group of servers in the server name field.

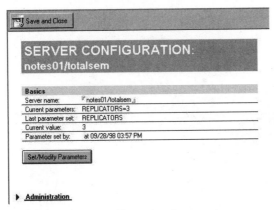

Figure 2.4 Server configuration document

If any current parameters are already set, in the case where you are editing an existing document, they will be shown in the Current parameters field. You can also see information about the last parameter that was set in the document using the Last parameter set, Current value, and Parameter set by fields.

To insert or change a parameter, click the Set/Modify Parameters button to display the Server Configuration Parameters dialog box shown in Figure 2.5. Server Configuration document information is available in the Configurations view of the NAB.

Use the drop-down arrow next to the Item field to select a parameter. A list of some of the parameters that you can edit using server configuration documents—and some parameters that you cannot edit using server configuration documents—is shown in Table 2.1. Use the Notes Administration Help database for more information regarding each of these parameters.

 NOTE
Many of the parameters available in this dialog box will be discussed later. Again, for those parameters not discussed, refer to the Notes Administration Help database.

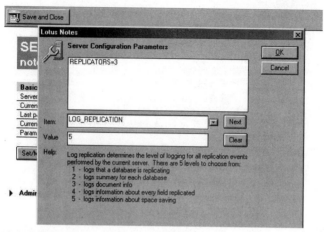

Figure 2.5   Server Configuration parameters dialog box

Table 2.1   Server Configuration Parameters

Parameters You Can Change in Server Configuration Documents	Parameters You Cannot Change in Server Configuration Documents
ADMINPINTERVAL	DOMAIN
ADMINPMODIFYPERSONDOCUMENTSAT	SERVER_CONSOLE_PASSWORD
AMGR_DOCUPDATEAGENTMININTERVAL	KITTYPE
AMGR_DOCUPDATEEVENTDELAY	ALLOW_PASSTHRU_CALLERS
AMGR_NEWMAILEVENTDELAY	ALLOW_PASSTHRU_CLIENTS
AMGR_NEWMAILAGENTMININTERVAL	ALLOW_PASSTHRU_TARGETS
AMGR_WEEKENDDAYS	ALLOW_PASSTHRU_ACCESS
BILLINGADDINWAKEUP	MAILSERVER
BILLINGADDINRUNTIME	MAILFILE

Parameters You Can Change in Server Configuration Documents	Parameters You Cannot Change in Server Configuration Documents
BILLINGCLASS	PORTS
BILLINGSUPPRESSTIME	ADMIN_ACCESS
DEFAULT_INDEX_LIFETIME_DAYS	CREATE_REPLICA_ACCESS
KILLPROCESS	SERVERNAME
LOG_AGENTMANAGER	SERVER_TITLE
LOG_MAILROUTING	TYPE
LOG_REPLICATION	NAMES
LOG_SESSIONS	ALLOW_ACCESS
LOG_TASKS	DENY_ACCESS
LOG_VIEW_EVENTS	CREATE_FILE_ACCESS
MAILCLUSTERFAILOVER	FORM
MAIL_LOG_TO_MISCEVENTS	ALL PARAMETERS BEGINNING WITH $
MAILDISABLEPRIORITY	
MAILLOWPRIORITYTIME	
MAILMAXTHREADS	
MAILTIMEOUT	
MEMORY_QUOTA	
NAME_CHANGE_EXPIRATION_DAYS	
NETWARENDSNAME	
NWNDSUSERID	
NWNDSPASSWORD	
NO_FORCE_ACTIVITY_LOGGING	

continues

Table 2.1   Continued.

Parameters You Can Change in Server Configuration Documents	Parameters You Cannot Change in Server Configuration Documents
NSF_BUFFER_POOL_SIZE	
PHONELOG	
REPL_ERROR_TOLERANCE	
REPL_PUSH_RETRIES	
REPLICATORS	
REPORTUSEMAIL	
SERVER_AVAILABILITY_THRESHOLD	
SERVER_MAXSESSIONS	
SERVER_MAXUSERS	
SERVER_RESTRICTED	
SERVER_SESSION_TIMEOUT	
SERVER_SHOW_PERFORMANCE	
SERVERPULLREPLICATION	
SHARED_MAIL	
SHOW_TASK_DETAIL	
SWAPPATH	
UPDATE_NO_FULLTEXT	
UPDATE_SUPPRESSION_LIMIT	
UPDATE_SUPPRESSION_TIME	
UPDATERS	

After you select a parameter, help information for that parameter will be displayed in the Help area of the dialog box. This information will help you understand the type of value to place in the Value field and whether the configuration will require you to shut down and restart the server. Place the appropriate value in the Value field.

At this point, either you can choose OK to save the parameter into the Server Configuration document, or you can click the Next button to add another parameter and value.

Server Connection Documents

Server connection documents in the Public NAB define the connection method and path between servers in your domain and in adjacent domains. In the Personal NAB, the connection documents define the path and connection information between a client and a server—and these are discussed later. Connection documents also set the replication and routing schedules for the servers in your domain. Connection documents can define paths to other servers using the following methods:

- Local Area Network connection documents make connections using LAN ports and protocols, such as TCP/IP or IPX/SPX. This type of connection is also used for a bridged or routed WAN connection, as Notes does not recognize that level of the network function.

- Dial-up Modem connection documents make connections to other servers using a communications port, such as COM1. You must provide dialing information for this type of connection.

- Passthru Server connection documents make connections to other servers, using a type of server designated as a Passthru, or stepping stone, server. Passthru is discussed in Chapter 5, "Notes Replication," and in Chapter 6, "Administration Tools and Tasks."

- Remote LAN Service connection documents make connections using a remote LAN service, such as *Microsoft's Remote Access Service* (RAS) or AppleTalk Remote access. To create remote LAN service documents, provide a username and other connection information for the LAN service.

NOTE
Different versions of Notes may include additional connection methods. This list, however, gives a representative sample of the available connection methods in Notes R4.X.

To create a connection document, select the Connections view and click the Add Connection action button. Notes also creates connection documents automatically, although they are not enabled, when you register and configure additional servers in your domain. Figure 2.6 shows an example of a Server Connection document. You can view Server Connection document information in the Connections view in the NAB.

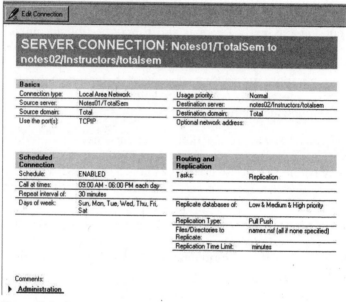

Figure 2.6   Server Connection document

In the Basics section of the Connection document, first select the connection type from the available list. This selection changes the other available options in the document. For our discussion, we assume that you select LAN connection type. Type the hierarchical names of the source, or initiating server, and the destination, or remote, server. You will also type the domain for these servers. Use the Choose ports button to select a port for communication, such as COM1 or TCP/IP. If necessary, you can give additional information necessary to reach the server in the Optional Network Address field, such as an IP address.

The Scheduled Connection section enables the administrator to determine the calling days, times, and intervals for the connection.

You may choose to have multiple connections to a particular server that do different tasks at various times. The schedule can either be enabled or disabled using the Schedule field. You may then choose Call at times. This field determines when the source server should try to make a connection to the destination server. You can place a single time, a range of times, or both in this field. The repeat interval field determines how often the server will attempt to call within the range of times defined in the Call at times field.

In the Routing and Replication section, first choose the tasks that this connection document should perform. The main choices are either replication or mail routing. The choice in this field changes the other options in the section. If you choose Mail Routing, you have the fields *Route at once if* and *Routing Cost*. These fields are discussed in detail in Chapter 4, "Notes Mail." If you choose Replication, you must complete the additional fields of *Replicate databases of what priority*, *Replication type*, *Files/Directories to replicate*, and *Replication time limit*. These fields are discussed in detail in Chapter 5, "Notes Replication."

Domain Documents

Domain documents specify and define the name, location, and type of access to adjacent and nonadjacent Notes domains and non-Notes domains. This type of document is used for routing mail among other domains and is discussed in detail in Chapter 7, "Advanced Configuration and Setup."

Mail-In Database Documents

A Mail-in database document enables any database to receive mail messages. Some Notes applications are designed to collect mailed-in forms in a database. If a database other than a user's mail file will be used to collect this data, it must have a Mail-in database document to specify how mail will be addressed to it and received by it. By default, for example, Notes creates a statistics collection Mail-in database and places a Mail-in database document in the NAB for this database when you create and configure Notes servers.

To create a Mail-in database document, open the Mail-In Databases view and click the Add Mail-In Database action button. This button displays the Mail-in database document shown in Figure 2.7.

First, complete the Mail-in name for the database in the Basics section of the document. This name will be used in formulas and mail messages to direct mail to this database. You may want to add a description to make the document easier to maintain. In the Location section of the document, you must complete information about the domain, server, and filename of the database. The domain should be the mail domain in which the database exists. The server should be the hierarchical name of the server on which the database is stored. The file name should be the file name and path for the database, relative to the Notes data directory.

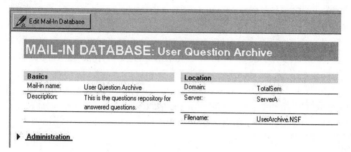

Figure 2.7   Mail-in database document

Server Program documents

Server programs can be used to enable administrators to automate complex administration tasks. Examples of these tasks include compacting databases and updating full-text indexes. Server programs can be run at the server console (see Chapter 6, "Administration Tools and Tasks"), or by creating Server program documents to schedule the program. Some programs also run automatically at server startup, whether or notif they have been added to the SERVERTASKS= or SERVERTASKSAT= lines in the NOTES.INI file. Examples of Server programs that can be run using a Server Program document are listed in Table 2.2.

To create a Server Program document to run one of the programs at a scheduled time, open the Programs view and click the Add Program action button. This button displays a Program document, as shown in Figure 2.8.

Table 2.2 Server Programs

Program	Load Program Using	Description
Administration Process	ADMINP	Automates a variety of administrative tasks
Agent manager	AMGR	Runs agents on one or more databases
Cataloger	CATALOG	Updates the Database Catalog
Chronos	CHRONOS	Updates full-text indexes that are marked to be updated hourly, daily, or weekly
Collector	COLLECT	Collects statistics for multiple servers
Database compactor	COMPACT	Compacts all databases on the server to remove unused white space
Database FIXUP	FIXUP	Locates and fixes corrupted databases
Designer	DESIGN	Updates all databases to reflect changes to templates
Event	EVENT	Monitors events on a server
Indexer	UPDALL	Updates all changed views and/or full-text indexes for all databases
Object store manager	OBJECT	Performs maintenance activities on databases and mail files that use shared mail
Replicator	REPLICA	Replicates databases with other servers
Reporter	REPORT	Reports statistics for a server
Router	ROUTER	Routes mail to other servers

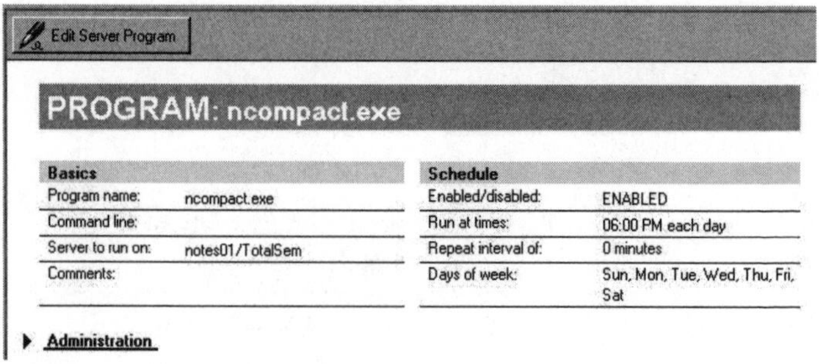

Figure 2.8 Server Program document

> **NOTE**
> The program names may have prefixes when run in certain oper-
> ating systems. If you are running your Notes server on a Windows
> NT server, for example, you might want to add an *N* to the begin-
> ning of each of the program names when adding them to a Program
> document. The programs will run without the added prefix in most envi-
> ronments, though.

Type the name of the program in the Program name field. Configure the program by filling in the various fields. If the program
has any command line arguments, those arguments can be placed
in the Command Line field. Type or select the hierarchical name of
the server on which the program should run. In the Schedule section, choose whether the schedule should be enabled or disabled.
You will then the select the times at which the program should run.
Use military-style time, from 0 (midnight) to 23 (11 p.m.) in this
field. You may also use a range of times in this field, and if you do
so, you will want to specify a repeat interval in that field to select
how often the program should run within the time range. Finally,
select the days of the week on which this schedule should be used.

Server Document

The server document is one of the most important documents in
the Public NAB, because it is used to define and configure the
servers in your domain. Notes creates a server document automatically when you register and configure a Notes server. To create or
edit a server document, use the Servers view of the NAB. Open an
existing server document in edit mode or create a new server document by pressing the Add Server action button. Notes displays a
server document, as shown in Figure 2.9.

Edit Server

SERVER: Notes02/Instructors/TotalSem

Basics

Server name:	Notes02/Instructors/TotalSem	Server build number:	
Server title:		Administrators:	Michael Meyers/TotalSem, Admin
Domain name:	Total	Routing tasks:	Mail Routing
Cluster name:		Server's phone number(s):	
Master address book name:			

▶ **Server Location Information**
▶ **Network Configuration**
▶ **Proxy Configuration**
▶ **Security**
▶ **Restrictions**
▶ **Agent Manager**
▶ **Administration Process**
▶ **Web Retriever Administration**
▶ **Internet Port and Security Configuration**
▶ **HTTP Server**
▶ **LDAP Server**
▶ **NNTP Server**
▶ **Internet Message Transfer Agent (SMTP MTA)**
▶ **X.400 Message Transfer Agent (X.400 MTA)**
▶ **cc:Mail Message Transfer Agent (cc:Mail MTA)**
▶ **Contact**
▶ **Administration**

Figure 2.9   Server document

NOTE
Adding a Server document to the NAB does not accomplish all the necessary tasks for registering the Server in the domain. Servers need to have certified ID files, replicas of the Public NAB, and other information created for them to be fully registered servers in the domain.

The *Basics* section of the server document contains the server name, the server title, the domain to which the server belongs, the server's build number, the server's phone number, the server's routing tasks, and the administrators for the server. The server name should be the fully distinguished hierarchical name. The title field is just informational. The domain name is required to define the Notes domain to which the server belongs. This field would be modified only if the server were being moved to another domain. All Notes administrators who require the ability to access the remote console should be in the administrator's field. You can change the server document as well in the administrator's field. The phone number field is only for information, used when users need to call into the server itself. The routing tasks field describes the types of mail routing that the server permits.

The *Server Location Information* section contains information for allowing the server to make telephone calls. The fields include the prefix to dial for an outside line, any international prefix that might be needed from this location, the local country code, any long distance prefix that might be needed at this location, and the local area code. This section also contains fields for calling card configuration. The dialing rules button can be used only if the server has a port enabled that can make telephone calls. The Remote LAN timeout field lets the administrator enable the server to terminate the call after a certain period of inactivity.

The *Network Configuration* section, shown in Figure 2.10, enables the administrator to specify the ports in use on this server. The port is directly related to the protocol enabled in the operating system. Use the TCP/IP port for the TCP/IP protocol stack, for example. The Notes Network field defines the server's Notes Named Network, as described in Chapter 1, "Installation and Configuration." The Net Address field for TCP/IP can be either the server's IP address or the server's DNS name. Each server can have multiple ports enabled and can be a member of multiple NNN.

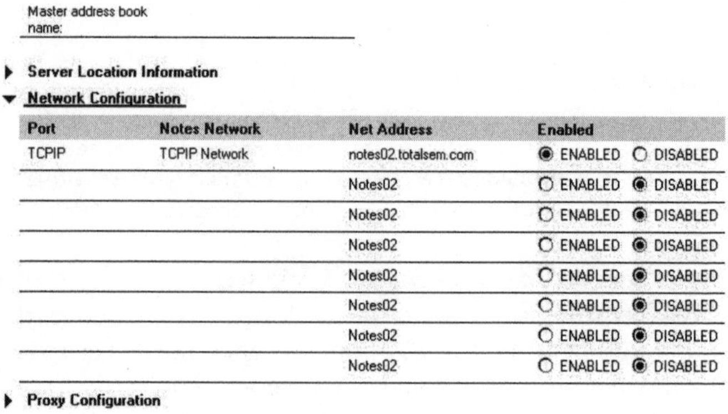

Master address book
name:

▶ **Server Location Information**
▼ **Network Configuration**

Port	Notes Network	Net Address	Enabled	
TCPIP	TCPIP Network	notes02.totalsem.com	⦿ ENABLED	○ DISABLED
		Notes02	○ ENABLED	⦿ DISABLED
		Notes02	○ ENABLED	⦿ DISABLED
		Notes02	○ ENABLED	⦿ DISABLED
		Notes02	○ ENABLED	⦿ DISABLED
		Notes02	○ ENABLED	⦿ DISABLED
		Notes02	○ ENABLED	⦿ DISABLED
		Notes02	○ ENABLED	⦿ DISABLED

▶ **Proxy Configuration**

Figure 2.10 Network Configuration section

The *Security Settings* section, shown in Figure 2.11, sets a few of the security parameters for the server. In this section, you have the ability to determine whether the server will compare Notes public keys of users with servers that attempt to authenticate with the

Notes public keys stored in the Public NAB. This option is intended to prevent users from using fake user ID files. This section also gives to the administrator the ability to allow Anonymous Notes connections. This feature enables any Notes ID to authenticate with the server, regardless of the certificates contained by the Notes ID.

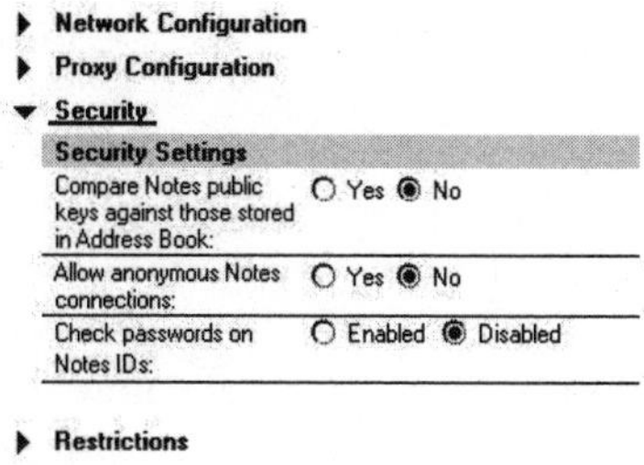

Figure 2.11   Security Settings section

The *Restrictions* section, shown in Figure 2.12, enables the administrator to restrict access to the server for general use, for Passthru use, and for creating new and replica databases. You can choose to allow access to this server only for users who are listed in the NAB, so that no users from other domains can gain access.

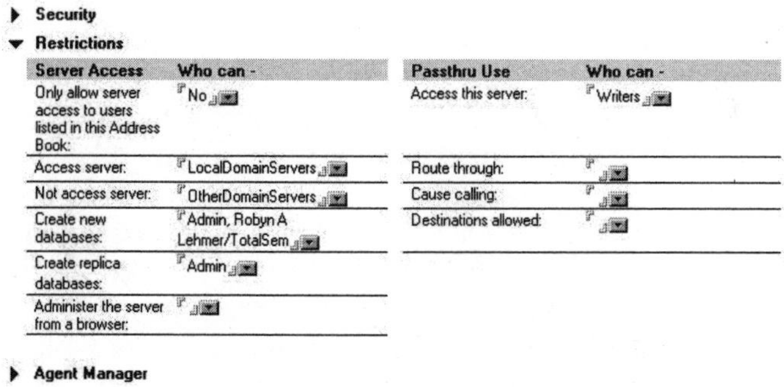

Figure 2.12   Restrictions section

You can complete the Access this server field to include users and servers that should have access to this server. If you leave this field blank, all users, servers, and groups have access to the server. To deny access to the server to specific users, servers, and groups, place their names in the Not access server field. This field can be populated with a Deny Access group, for example.

Place any users who should have the ability to create new databases on the server in the Create new databases field. If this field is left blank, all users, servers, and groups can create new Notes databases on the server. Place any users who should have the ability to create new replicas of databases on the server in the Create replica databases field. If this field is left blank, no users, servers, or groups can create new replicas on the server.

Place any users, servers, and groups that should be able to access the server via a Passthru server in the Access this server through Passthru field. If this field is left blank, no users, servers, or groups may access this server using a Passthru server. In the Route through field, place the names of all users, servers, and groups that should be able to use this Notes server as a Passthru server to access other Notes servers. If this field is left blank, no users, servers, or groups can use this server as a Passthru server. Use the Passthru Cause calling field similarly. Again, if the field is blank, no one can use this server for Passthru. Finally, you can specify which destination servers may be accessed by users who are using this server as their Passthru server. Passthru is discussed in more detail in Chapter 5, "Notes Replication," and in Chapter 6, "Administration Tools and Tasks."

The *Agent Manager* section of the server document, shown in Figure 2.13, enables an administrator to define which users can run agents on the server, and when those agents can be run. An agent is a document created by a user or a designer to automate manual database functions. An agent is similar to a macro in other applications. Add the users, servers, and groups that should have the capacity to run personal, restricted, and unrestricted agents on the server to the appropriate fields in the agent restrictions section. If the Run Personal Agents field is left blank, all users can run personal agents. If the Restricted and Unrestricted Lotus Script Agents fields are left blank, no one can run these types of agents on the server.

Note that you can set the agent manager to have different settings during the day and night. Both options enable the administrator to define starting and ending times for agents to run on the server. This option enables the administrator to forbid agents to run when other scheduled programs may be running. The administrator can choose how many agents can be running on the server at the same time. This option enables the administrator to control the performance of the server. Too many agents running simultaneously on the server may take away from necessary tasks, such as mail routing. Similarly, the administrator can choose the maximum execution time for Lotus Script agents. Again, if they are allowed to run for an unlimited amount of time, other necessary resources may be overused, and other performance may suffer. Finally, this section enables the administrator to define the amount of the server's time, in a percentage, that can be spent running agents.

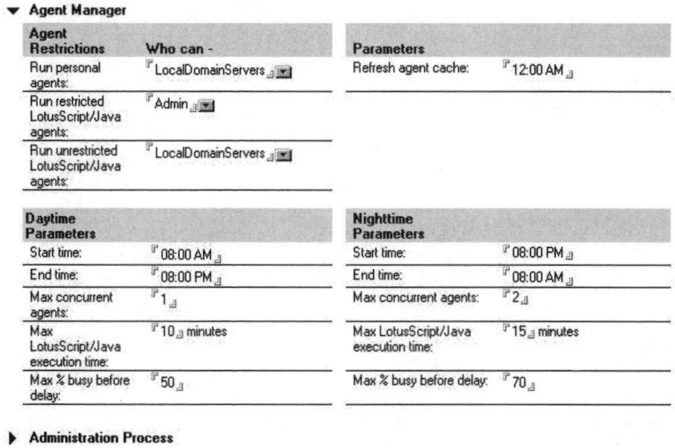

▼ Agent Manager

Agent Restrictions	Who can -		Parameters	
Run personal agents:	LocalDomainServers		Refresh agent cache:	12:00 AM
Run restricted LotusScript/Java agents:	Admin			
Run unrestricted LotusScript/Java agents:	LocalDomainServers			

Daytime Parameters		Nighttime Parameters	
Start time:	08:00 AM	Start time:	08:00 PM
End time:	08:00 PM	End time:	08:00 AM
Max concurrent agents:	1	Max concurrent agents:	2
Max LotusScript/Java execution time:	10 minutes	Max LotusScript/Java execution time:	15 minutes
Max % busy before delay:	50	Max % busy before delay:	70

▶ Administration Process

Figure 2.13   Agent Manager section

The *Administration Process* (AdminP) section, shown in Figure 2.14, enables the administrator to control the way that AdminP runs on the current server. AdminP, which will be discussed in depth in Chapter 6, "Administration Tools and Tasks," assists the administrator with routine administrative tasks, such as renaming or recertifying a user. These options enable the administrator to define the interval to carry out requests. If there is no value in this field, Notes looks for an ADMINPINTERVAL setting in the NOTES.INI or uses the default 60 minutes as the interval. This section also enables the administrator to specify the time that AdminP should complete requests that are scheduled for once a day. Finally, requests that are of a delayed type are set to run at the time specified in this section.

EXAM TIP
This section has described the majority of the Notes server document that relates to the default functioning of the server. If you have installed additional options or add-in server tools, you may see additional sections in the server document. Additionally, when you are using Notes R4.5 or 4.6, you may see additional sections in the server

document. These additional sections will not be tested on the System Administration I exam.

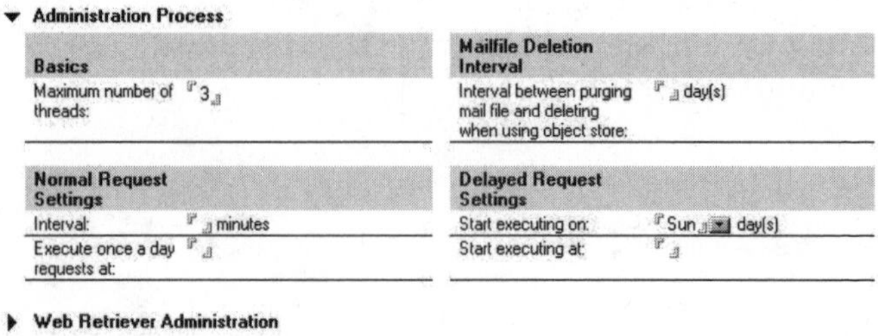

Figure 2.14   Administration Process section

User Setup Profile Document

When you registered a user in Chapter 1, "Installation and Configuration," you may have noticed the field on the Register Person dialog box called Profile. The User profile document enables an administrator to set a variety of workstation defaults for multiple users, including Internet settings, Passthru settings, and the databases that appear on their workspaces. To create a User profile, open the NAB to the Server . . . Setup Profiles view. Click the Add Setup Profile button to display the New Setup Profile form, shown in Figure 2.15.

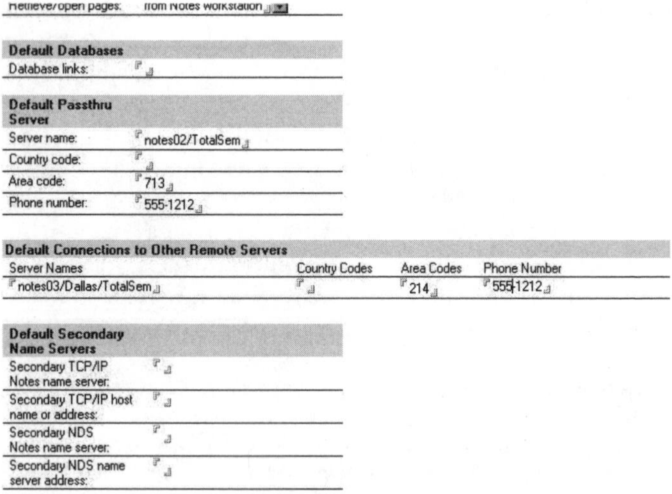

Figure 2.15   User Setup profile

First, type a name for the profile. This name should be descriptive of the users for whom the profile is issued. Complete information describing the users' Internet and Passthru defaults. Complete any other necessary information and then save the document. Additionally, the User Setup Profile can be used to define the users' connections to other servers.

Personal Name and Address Book

While the Public NAB configures the users, servers, groups, and connections for an entire domain, the Personal NAB contains configuration and connection information for a single user. When a user's workstation is configured, Notes creates a Personal NAB for them using the PERNAMES.NTF template. The Personal NAB will be named with the user's last name and will be stored in their local Notes data directory.

The Personal NAB contains documents that are similar to the documents of the same names in the Public NAB. The Personal NAB contains the following documents:

- *Person documents.* In the Personal NAB, Person documents are used to maintain a list of contacts.

 NOTE
In Notes R4.6, Person documents in the Personal NAB are changed to Business Cards.

- *Group documents.* In the Personal NAB, Group documents are used to maintain personal mailing lists.

- *Certifier documents.* The certifier documents in the personal NAB are used to identify the certificates held by a particular user. These documents are used heavily during cross-certification and are described in Chapter 7, "Advanced Configuration and Setup."

- *Connection documents.* The Connection documents in the Personal NAB define methods of connecting the user's workstation to the appropriate servers. As you will see in Chapter 6, "Administration Tools and Tasks," these documents are used extensively with Passthru and other remote connections.

- *Location documents.* Location documents give users the ability to store different default settings for mail file location, connection types, and ports for users who connect from different places, such as home, the office, or hotels. The default location documents included with the workstation are Home (modem), Internet, Island (disconnected), Office (network), and Travel. An example of a Home location document is shown in Figure 2.16.

LOCATION: Home (Modem)

Basics

Location type:	Dialup Modem	Prompt for time/date/phone:	No
Location name:	Home (Modem)		

Internet Browser

		Servers	
Internet browser:	Notes with Internet Explorer	Home/mail server:	notes01/TotalSem
Browse:	direct from Notes workstation	Passthru server:	

Ports

Ports to use: [X] TCPIP

Phone Dialing

Prefix for outside line:		Calling card access number:	
International prefix:		Calling card number or extension suffix:	
Country code at this location:		Dialing Rules...	
Long distance prefix:	1		
Area code at this location:	713		

Mail

		Replication	
Mail system:	Notes and Internet	Schedule:	Disabled
Mail file location:	Local		
Mail file:	mail\sjerniga		
Notes mail domain:	Total		
Recipient name type-ahead:	Personal Address Book Only		

Figure 2.16   Office location document

One way that users can switch between locations is by using the Location button on the status bar of the workstation, as shown in Figure 2.17. Location documents are also discussed as they relate to user Passthru in Chapter 6, "Administration Tools and Tasks."

Figure 2.17   Changing location

NOTE
When users complain of problems connecting to their mail server
or other Notes resources but they say that they can access non-
Notes resources, their Location is one of the first things to check.
If Notes thinks that it is in a disconnected state, for example, it will not
even try to connect to a Notes server.

Review Questions

1. Which default groups are created in the NAB?

   a. Default and LocalDomainServers

   b. Everyone and OtherDomainServers

   c. LocalDomainServers and OtherDomainServers

   d. Default and Everyone

2. Emily created a group in the NAB called Terminations. When she goes to the Groups view after saving it, that group is not visible. Why?

   a. Terminations is a default group and cannot be saved as a new group. It can be only edited.

   b. She created it as a Mail only group, so it can be seen only in the Address dialog box for messages.

   c. She created it as a Deny List only group, so it can be seen only in the Address dialog box for messages.

   d. She created it as a Deny List only group, so it can be seen only in the Deny Access Groups view of the NAB.

3. You cannot put another group inside of a group—only people and servers.

   a. True

   b. False

4. Susan created a group to contain all the employees of her organization. Suddenly, when she is adding the new users she has created, it will not allow her to save. What might be the explanation?

   a. The address book is corrupt.

   b. You cannot add users to groups after creating them; instead, you must recreate the group.

   c. The group is too large. Create a second group and nest it.

   d. The users have not configured their workstations yet and cannot be placed in the group.

5. Creating a Person document is the same as choosing Register Person from the Server Administration Panel.

   a. True

   b. False

6. When you edit the Person document, you can change the User Name if the person needs to be renamed.

   a. True

   b. False

7. Maggie is a developer in your organization. When you created her user ID, you gave her a Notes Desktop license. She is unable to do development work. How do you fix this problem?

   a. You cannot. She installed the wrong software components.

   b. You can change her license type in her Person document in the NAB.

   c. You must re-register her and create a new user ID with the correct license.

   d. You can choose Upgrade License from the Actions menu in the NAB.

8. When you create a Server Configuration document, you are changing settings where?

   a. You are changing settings in the Server document of the NAB.

   b. You are changing settings in the server's NOTES.INI.

   c. You are changing settings in the local NOTES.INI.

   d. You are changing settings in the PERNAMES.NTF.

9. All changes made using a Server Configuration document take effect immediately.

   a. True

   b. False

10. All settings in the server's NOTES.INI can be changed using a server configuration document.

    a. True
    b. False

11. What two tasks can be scheduled using a Connection document?

    a. Replication and Server Compaction
    b. Mail Routing and Server Compaction
    c. Mail Routing and Backups
    d. Mail Routing and Replication

12. What options are available if you make a Remote LAN service type of Connection document?

    a. Microsoft RAS
    b. AppleTalk Remote Access
    c. NDS Remote
    d. IBM RAS

13. The only reason to create a Mail-in database document is to enable a user's mail file to accept mail.

    a. True
    b. False

14. Wayne wants to force the Cataloger task to run at midnight and again at noon. How can he do this procedure?

    a. Type LOAD CATALOG at the console at midnight and noon.
    b. Add a Program document to the NAB.
    c. Add SERVERTASKSAT12= to the NOTES.INI.
    d. The Cataloger only runs at 1 a.m., as set in the NOTES.INI.

15. You can use the Owners and Administrators field to give any user with Author access the ability to edit documents in the NAB.

    a. True
    b. False

16. If you leave the Create New Databases field blank in the Restrictions section of a server document, who can create new databases on the server?

    a. Everyone
    b. No one
    c. Administrators
    d. LocalDomainServers

17. If you leave the Create Replica Databases field blank in the Restrictions section of a server document, who can create new databases on the server?

    a. Everyone
    b. No one
    c. Administrators
    d. LocalDomainServers

18. If you leave the Access this server field blank, who can access the server?

    a. Everyone
    b. No one
    c. Administrators
    d. LocalDomainServers

19. If you leave the Access this server through Passthru field blank, who can access the server using Passthru?

    a. Everyone
    b. No one
    c. Administrators
    d. LocalDomainServers

20. If you place a user in both the Access this server field and the Not access server field, what happens?

    a. They can access the server.

    b. They cannot access the server.

    c. They can access the server if their name is placed individually in the Access this server.

    d. They cannot access the server if their name is in a group in either field, it must be placed in the fields individually.

Review Answers

1. c. LocalDomainServers and OtherDomainServers

2. d. She created it as a Deny List only group, so it can be seen only in the Deny Access Groups view.

3. b. False. You can nest up to six levels of groups.

4. c. The group is too large. Create a second group and nest it.

5. b. False. The user has no ID file and does not have any keys or certificates.

6. b. False. Use the Rename person action to change the user's name. Do not change the first, last, middle, or user name fields.

7. b. You can change her license type in her Person document in the NAB.

8. b. You are changing settings in the server's NOTES.INI.

9. b. False. Some settings require you to restart.

10. b. False. Some settings can be changed only in a server document or directly in the NOTES.INI.

11. d. Mail Routing and Replication

12. a and b. Microsoft RAS and AppleTalk Remote Access are the only types of remote access service available in Notes R4.

13. b. False. A user's mail file can accept mail by default and does not need a Mail-in database document.

14. a is possible, but b is the best answer. Add a Program document to the NAB.

15. a. True

16. a. Everyone

17. b. No one

18. a. Everyone

19. b. No one

20. b. The Not access server field takes precedence, and they cannot access the server.

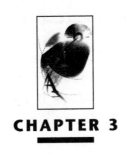

CHAPTER 3

Notes Security

One of the most important elements in configuring and maintaining a Notes environment is security. Administrators should understand the levels of security that are available and how and when to implement those levels. In addition, certain elements of the Notes environment have more security components, such as the ID files, the Public NAB, and Notes Mail. The key to unlocking the secured elements of a Notes environment is a user's name and identity, stored in their ID file. This chapter describes the user ID and how to use it to gain access to the Notes environment. In addition, you will learn about the levels of security in Notes—from gaining access to the server to gaining access to the forms and views in databases. Finally, you will learn about the additional security elements that can be applied to certain facets of the Notes environment.

Objectives

After reading this chapter, you should be able to answer questions based on the following objectives:

- Define the elements of a user or server ID file.
- Understand anti-spoofing and anti-guessing password security elements on the ID files.

- Understand the levels of Notes security.

- Understand the server access and authentication process.

- Understand how to create and use Directory links.

- Describe the levels of the *Access Control List* (ACL).

- Use groups and roles in the ACL.

- Describe the additional access control applied to the NAB.

- Describe Public Key/Private Key encryption.

- Describe Signing.

Notes Server and User ID Files

The Notes Server and User ID files contain all the information that a server or user needs to access and pass through the various levels of Notes security. Without an ID, however, access is denied to Notes servers or databases. This clearly important file contains its own security elements. In Chapter 1, "Installation and Configuration," we discussed creating an ID for a new user or server by using the registration process. This chapter builds on that knowledge by describing the elements that are contained in the ID file and how the file is protected.

Components of the Notes ID

The Notes ID file for a user or server contains all the information necessary to identify a Notes user or server to the Notes environment. This information includes not only the user's name but also the following additional items, as shown in Figure 3.1.

- *User Name*. The common name of the Notes user is stored in the ID file.

- *Notes License type*. When a user is created, the administrator assigns a type of license—either North American or International. The type of license cannot be changed after the ID is created.

- *Public and Private keys*. These randomly generated, hexadecimal keys uniquely identify the user. The keys contain an element of the certifier's key that was used to register the user. The keys are used in authentication, encryption, and signing. The Public key, as we saw in Chapter 2, "The Notes Name and Address Book," is also stored in the user's Person document in the Public NAB. The Public and Private keys are a mathematically related pair.

- *Encryption keys.* Encryption keys are secret keys generated by you or another user, which enable you to encrypt and decrypt fields on a form.

- *Certificates.* The original certificate granted to a hierarchical ID is stored in the ID file. Any additional certificates are stored in the NAB. If the ID file is flat, all certificates are stored in the ID file. Also, any flat certificates given to a hierarchical user are stored in the ID file.

- *Password.* Each ID file can contain a password for protecting the ID file itself.

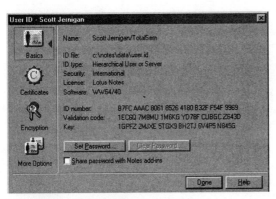

Figure 3.1   Components of an ID file

The user or server ID file is stored in the Notes data directory by default, after the user or server has been configured. Many users and administrators also maintain a backup copy of the ID files, in case of loss or corruption. Although an administrator could generate a new ID file for a user or server if necessary, this ID file would contain different public and private keys—and none of the secret encryption keys. Therefore, if the user encrypted any mail or other documents, no one could read them.

Security for the Notes User ID

Notes uses a password to protect the Notes ID files. When you registered users, servers, and certifiers, you specified the minimum password length required for each ID that you created. While it is possible to choose not to password-protect an ID file—by selecting zero as the minimum number of characters for the password—you would be opening up your Notes environment to security breaches. The password, which can be up to 63 characters, is used to encrypt the ID file. When a user types in the password correctly, the ID file is decrypted and can be used to access the server or other objects.

NOTE

By default, the password is only stored in the ID file and is only used to protect the ID file. In Notes R4.5 and higher, you also have the option of checking the password at the server to verify that the correct ID is being used—forcing users to change their passwords after a set period of time and to keep a log of the passwords to prevent reuse.

To change or set a password, the user will examine the ID file by choosing File . . . Tools . . . User ID. On the Basics panel, click the Set Password button to change the password for the ID file. Notes prompts you to decrypt the ID using the current password, before enabling the user to change the password. When the Set Password dialog box shown in Figure 3.2 is displayed, the user can type the new password. Remember that passwords in Notes *are* case sensitive. After typing and confirming the password, the password in the User ID file is reset.

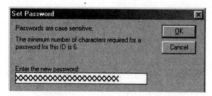

Figure 3.2 Set Password

Notes provides the capacity to set multiple passwords on a server or certifier ID file. This feature enables an administrator to require multiple administrators to be present to access a server or certifier ID. To set multiple passwords on a server or certifier ID file, use the Certifiers button in the Server Administration panel. For more information regarding multiple passwords, refer to the Notes Administration Help database.

Notes passwords are protected from password-guessing and password-capturing programs by two additional security features.

Password-capturing programs are deterred by the anti-spoofing mechanism of Notes passwords. This anti-spoofing mechanism is represented by the hieroglyphics pattern displayed on the password dialog box as users enter passwords. Typical password-capturing programs find it difficult to emulate these changing patterns of hieroglyphics. They can not therefore emulate the Notes login screen to capture the users' passwords. The time-delay feature of Notes passwords deters password-guessing programs. If a user types an incorrect password to a Notes ID file, Notes takes additional time to display the error message and give the user an additional opportunity to retype the password. The more times the password is typed incorrectly, the longer the interval. Notes increases the interval exponentially, making it difficult for password-guessing programs to try a large number of passwords in an attempt to break into the ID file.

Notes Security Overview

Security in a Lotus Notes environment actually begins with security outside the environment. To secure your server, applications, and data, the security should begin at a hardware level. First, place the servers and workstations in your environment in locked areas to prevent physical access. Second, fully implement the network operating system and the computer operating system security features to prevent access at the operating system level. These two levels of security are not necessarily the job of the Notes system administrator and will not be tested on the exam. They are, however, vital to the way in which you will implement Notes security measures. Whether someone gains physical or network access to your servers, they may be able to circumvent many of the Notes security measures.

After gaining access to the Notes ID, the user or server must pass through additional security layers before gaining access to the data in Notes. The next levels of security are Notes security measures. The following list outlines basic options for securing the Notes environment, as also shown in Figure 3.3:

- Authentication in the Notes organization
- Notes Server access lists
- Database directory links
- Database access control lists
- User roles to refine user's access
- Form and/or View access lists

- Document access through Reader and Author fields
- Field and mail security through encryption
- Mail and data verification through signing

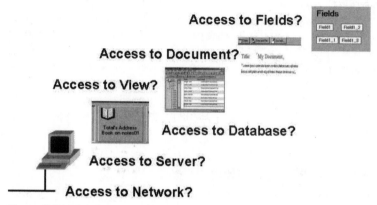

Figure 3.3   Notes security levels

Not all of these security levels are the sole responsibility of the Notes administrator. The administrator has primary responsibility for server access, using IDs and access lists, and using ACLs for database access. Below this level of security, the administrator shares the security responsibility with the developer of the database. The lower levels of security are discussed in this chapter only slightly, because they are not tested on the System Administration exams. You should, however, understand how all the layers of security work together.

Lotus Notes Security

The first two levels of security within Notes are access to the server through authentication and the server access control list. The server administrator creates and enforces these levels of security.

Notes Authentication

Authentication is the process by which a user or server, using the ID file, verifies their identity within the Notes Organization. Each Notes Organization has one or more certifiers. Each ID file within the Notes

Organization is stamped with a certifier. The authentication process compares the stamp on the user's or server's ID file with the stamp that is part of the Notes Organization, determining whether it has any shared certifier stamp in common. The following steps are used in the process of authentication, as shown in Figure 3.4:

1. The server or user requesting access generates a random number. The requesting server or user sends to the destination server a package that contains this random number, their name, their public key, and their certificates.

2. The server takes the random number and signs it, using the signing methods that will be described later. The server then sends this information back to the requesting server or user.

3. When the requesting user or server receives the signed random number, it verifies that the signature is correct and that the random number matches the original random number. Because signing uses the public key for verification, only users with the correct public key can verify that the correct signature has been appended.

4. The server then repeats the entire process in reverse, to prove to the requesting server or user that the server is not an impostor.

Figure 3.4 Authentication

Authentication is always a two-way process of verifying the identity of both the requester and the requested servers or users. Authentication is based on the trusted information (the common certificate) that the certifier gives to both parties.

Authentication is successful when these statements are true:

- Hierarchical IDs must have a common certifier in their tree (common ancestor).

- Flat IDs must have some common certificate.

- Hierarchical IDs gaining access to flat resources must have some common certificate.

- If hierarchical IDs do not have a common ancestor, they must use the process of cross-certification to gain a common certificate, as described in Chapter 7, "Advanced Configuration and Setup."

Notes Server Access

After using the ID to verify identity and gain access to the environment, the users or servers that are trying to access a database must also gain access to the server on which the database is stored. Each server's configuration document contains a *server access list*, which is a list of users who are allowed (or not allowed) to access the server. The user must be allowed to gain access to the server based on the entries in this list. The server document contains two sections, as discussed in Chapter 2, "The Notes Name and Address Book," that define the users or servers that have the ability to access the server. These sections are shown in Figure 3.5.

The Security Settings section sets a few of the security parameters for the server. In this section, you have the ability to determine whether the server will compare Notes Public keys of users and servers that attempt to authenticate with the Notes Public keys stored in the Public NAB. This option is intended to prevent users from using fake user ID files. If an administrator chooses not to compare public keys, the authentication process described earlier is much more limited. This section also grants to the administrator the ability to enable Anonymous Notes connections—and enables any Notes ID to authenticate with the server, regardless of the certificates contained in the Notes ID.

▼ Security

Security Settings		
Compare Notes public keys against those stored in Address Book:	○ Yes ● No	
Allow anonymous Notes connections:	○ Yes ● No	
Check passwords on Notes IDs:	○ Enabled ● Disabled	

▼ Restrictions

Server Access	Who can -	Passthru Use	Who can -
Only allow server access to users listed in this Address Book:	No	Access this server:	
Access server:		Route through:	
Not access server:		Cause calling:	
Create new databases:		Destinations allowed:	
Create replica databases:			
Administer the server from a browser:			

Figure 3.5   Security and Restrictions sections of the Server document

NOTE

If you allow anonymous connections, you must also provide for anonymous users to access the databases. You can create an entry in the databases' ACLs called Anonymous. Users who do not authenticate will use the level of access determined by this entry. If you do not create an anonymous entry, anonymous users will be granted the default access to the database.

The Restrictions section is actually where the administrator has the opportunity to restrict access to the server—for general use, for Passthru use, and for creating new and replica databases. You can choose to enable access to this server only for users who are listed in the NAB, so that no users from other domains can gain access. You can complete the Access this server field to select users and servers that should have access to this server. If you leave this field blank, *all* users, servers, and groups have access to the server. A more secure method, however, is to place groups of servers and users in this field who should have explicit rights to access. This field is also called the server access list.

To deny access to the server to specific users, servers, and groups, place their names in the Not access server field. This field can be populated with a Deny Access group, for example, and overrides the access list. If a user is in both the Access this server field and the Not access this server field, they will not be able to access the server.

Place any users who should have the ability to place or create new databases on the server in the Create new databases field. If this field is left blank, *all* users, servers, and groups can create new Notes databases on the server. Place any users who should have the ability to create new replicas of databases on the server in the Create replica databases field. If this field is left blank, *no* users, servers, or groups can create new replicas on the server.

EXAM TIP

For the System Administration I exam, it is important to note the difference between these two fields and what occurs when each field is left blank.

Place any users, servers, and groups who should be able to access the server via a Passthru server in the Access this server through Passthru field. If this field is left blank, no users, servers, or groups may access this server using a Passthru server. In the Route through field, place the names of all users, servers, and groups who should be able to use this Notes server as a Passthru server to access other Notes servers. If this field is left blank, no users, servers, or groups can use the server as a Passthru server. Use the Passthru Cause calling field similarly. Again, if the field is blank, no one can use this server for Passthru. Finally, you can specify which destination server's users who are using this server as their Passthru server may access. Passthru is discussed in more detail in Chapter 5, "Notes Replication," and in Chapter 6, "Administration Tools and Tasks."

Notes Database Access

After gaining access to the server, users or other servers must then gain access to the databases that contain the data they need. The security elements for database access include Notes Directory links and Access Control Lists for the databases.

NOTES DIRECTORY LINKS

Databases can be stored in three places on Notes servers:

- Directly at the root of the default \NOTES\DATA directory
- In a subdirectory of the \NOTES\DATA directory, such as \NOTES\DATA\ MAIL
- In a directory elsewhere on the system, linked to the \NOTES\DATA directory using a directory link or a database link

To use a directory link, create a directory on the server that is not in the Notes data directory. Then, place a text file in the Notes data directory that contains the path to the remote directory. The text file should be named with the name of the directory and a .DIR extension. When users look in the Open Database dialog box, they will see the name of the directory as though it were stored in the Notes data directory.

Create database links in a similar manner but use the name of the database instead of the name of the directory when naming the text file. Users will see the database name in the Open Database dialog box.

To add security to the directory or database links, add the hierarchical names of users or group names that should have access to the directory or database. If a user's name is not in the directory or database link file, they will be able to see the directory or database but will not be able to gain access.

NOTE

In Notes R4.5 and higher, Lotus has made creating directory and database links much easier and more automated by using the Server Administration panel. To create a database or directory link in Notes R4.5 or higher, open the Server Administration panel and click on the Servers button. Choose Directories and Links from the drop-down list. Create a new link by typing in the file that you are linking, and indicate the path to the directory or database. Also choose the users who should have access, as shown in Figure 3.6.

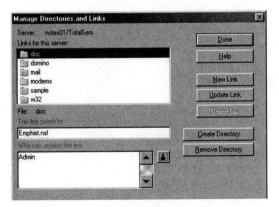

Figure 3.6   Creating a link in Notes 4.5X

NOTES ACCESS CONTROL LISTS (ACL)

One of the most important layers of security in Notes is the ACL. To determine who can gain access to a particular database and to determine what options the users or servers will have within that database, Notes provides an ACL for each database. The ACL determines the following three items of security:

- The users and servers that can access a database
- Definitions of functions that each user, server, or group can perform
- The ACL also enables a designer to add roles for users, servers, and groups, to refine access to the database

To view the ACL and its options, right-click on the database icon and choose Access Control, or choose File . . . Database . . . Access Control from the menus within any particular database. Now you will see the ACL dialog box, as shown in Figure 3.7. You can also view the ACL by choosing File . . . Database . . . Access Control . . . while a database is selected.

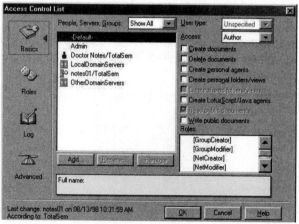

Figure 3.7   ACL dialog box

In the ACL dialog box, you have four panels: Basics, Roles, Log, and Advanced. Figure 3.7 shows the dialog box with the Basics panel displayed.

In the Basics panel, the center of the dialog box displays the levels of access for each user, server, and group, including the Default access level. The first item listed in this box should be the word *default*. The default access will apply for any users, servers, or groups that are not explicitly listed in the ACL. In most cases, for the tightest security, the default ACL should be set to No Access to prevent any user from inadvertently gaining access. The two default groups mentioned in Chapter 2, "The Notes Name and Address Book", LocalDomain-Servers and OtherDomainServers, should always be listed in the ACL with an appropriate level of access. In addition, other users, groups, and servers should be listed as appropriate. Please note that in most cases, it is desirable for groups rather than users to be listed in the ACL—for ease of administration. On the right side of the first panel, you will see the type and level of access for the users, groups, and servers that were listed in the center. In addition, you will see the additional rights that can be given with each level of access. Those rights are described in the following sections.

The Roles panel enables the administrator to refine access to certain design elements inside the database by creating and using user roles. User roles are described later in this chapter.

The Log panel shows all additions and modifications to the ACL of this database. The log shows the name of the user who made the ACL change, the time and date of the ACL change, and what change was made. The Log panel is shown in Figure 3.8.

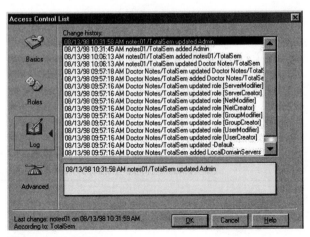

Figure 3.8 ACL Log

The Advanced panel enables an administrator to determine the administration server for the database, as described in Chapter 6, "Administration Tools and Tasks." The administrator can also use this panel to enforce a consistent level of security across all replicas of the database, which prevents users from making local replicas that they can change. You also can maintain data consistency and accuracy across different servers. The Advanced panel is shown in Figure 3.9.

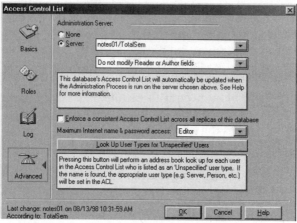

Figure 3.9 ACL advanced options

The ACL has seven basic levels of security that can be refined through assigning additional options. These levels of security are all set on this panel of the dialog box. To assign a user or server to a level of the ACL, click the Add . . . button at the bottom of the Basics panel of the dialog box. You then have the option whether to type the name of the user, server, or group, or to choose a name from the Public NAB. You may also remove or rename users, servers, and groups using the other buttons at the bottom of the dialog box. If a user is not listed specifically in the ACL, either individually or within a group, they will be given the access level of the —Default— group. The —Default— group, which cannot be renamed or removed, is given Designer access by default when a new database is created. The user who creates the database is also added to the ACL automatically—with Manager access—when the database is created. Two other groups, LocalDomainServers and OtherDomainServers, are also added to the ACL by default. These groups are standard

Notes groups that are used in all Notes environments to determine the access level of the servers that house and replicate the databases. Both of these groups have Manager access when the database is created.

 NOTE

I generally recommend choosing a name rather than typing it in, to ensure that the hierarchical name is correct. I also recommend using groups in the ACL as often as possible. This feature enables an administrator to change access for users by changing group membership, rather than having to open each database and change the ACL manually.

After choosing a user, group, or server, choose the appropriate user type from the drop-down list. Then, choose the Access level and options from the drop-down list and the check boxes. The following paragraphs describe each level of security and the options that are available for refining them.

No Access

The lowest level of access to a database is No Access. *No Access* prevents a user or group not only from opening a database but also from adding the database to their workspace. This level of security is often recommended for the —Default— access to the database, so that no one can gain access to data inadvertently. The only additional options available with No Access are the capacity to Read public documents and to Write public documents, as shown in Figure 3.10. None of the other options are selected or available. With No Access, a server will not be able to replicate documents.

Figure 3.10 No Access

 NOTE

Public access enables users with limited access to create or view certain documents or forms. These rights are available beginning in R4.5 only. Previous to R4.5, there were no additional options available for users with No Access. Public Access is not tested on the System Administration I exam.

Public access to documents, forms, and folders can be granted on the Form, Document, or Folder properties InfoBoxes on the Security tab. An example of Public access in use in Notes is the capacity for users to grant other users rights to their calendars or mail files. These rights are granted by giving users with No Access the ability to read and/or write public documents.

Depositor Access

The next level of access to a database is Depositor access. *Depositor access* enables a user to add the database to the workspace and open it. In addition, depositors can create documents in the database. They cannot, however, read or edit any documents in the database. The views in the database will appear empty to a user with Depositor access. Depositor access is useful for mail-in databases, surveys, suggestion boxes, or evaluations. The users can create an evaluation for an instructor, for example, but will not be able to see or change any of the evaluations. The Create documents option is checked by default and cannot be changed. No other options (except for the read and write public documents options in R4.5 and higher) are available, as shown in Figure 3.11.

Figure 3.11 Depositor access

Reader Access

You can also assign Reader access to a database. Users with *Reader access* to a database can view documents in the database but cannot create or edit database documents. This type of access is useful for informational databases, such as creating a corporate directory or a documentation database. The additional options for users with Reader access are Create personal agents, Create personal folders/ views, and Create LotusScript agents, as shown in Figure 3.12. The other options are all unchecked and unchangeable, with the exception of Read public documents in R4.5 and higher, which is checked and unchangeable. With Reader access, a server will not be able to replicate documents.

Figure 3.12   Reader access

Author Access

The next level of access to a database is Author access. Users with *Author access* to a database have the ability to read documents in the database, create documents in the database, and edit the documents they create. Author level access is the most common level of access to discussion and other databases, because it prevents Replication and Save Conflicts. The additional options available for users with Author access are Create documents, Delete documents, Create personal agents, Create personal folders/views, and Create LotusScript agents, as shown in Figure 3.13. In R4.5 or higher, the Read public documents option is checked and unavailable for change, and the Write public documents option is available for change. With author access, a server can replicate only new documents—although it is usually better to give servers Editor access.

Figure 3.13   Author access

Editor Access

Users with *Editor access* to a database start with the same level of access as users with Author access: the ability to read, create, and edit their own documents. Editor access also grants the ability to edit all other documents in the database. Usually, you should give only one or two users this level of access to a database, to prevent Replication and Save conflicts. The Create documents option is checked by default, as shown in Figure 3.14, and is not available for change. In R4.5, this statement is also true of both the Read public documents and Write public documents options. All other options are available for modification for users with Editor access—the first level of access that can grant the ability to create shared folders and views. Servers with Editor access can replicate new and changed documents.

Figure 3.14   Editor access

Designer Access

Designer access gives a user all the rights enabled by Editor access, including reading, creating, and editing all documents in the database. Designers have the additional ability to create and modify all design elements in the database, including forms, views, navigators, and others. Designers have the ability to create and modify full text indexes, About and Using this database documents, and database icons. The Create documents, Create personal agents, Create personal folders/views, and Create shared folders/views options are granted to users with Designer access by default and cannot be changed. In R4.5, this statement is also true of both the Read public documents and Write public documents options. The options for Delete documents and Create LotusScript agents are available to be modified, as shown in Figure 3.15. Servers with Designer access can replicate all new and changed documents and design elements.

Figure 3.15   Designer access

Manager Access

Users with *Manager access* are granted all the abilities of users with Designer level access, and they are the only access level of users given the ability to change the ACL of the database. Managers can also encrypt a database, delete a database, and determine the replication settings for a database. All options are granted by default to Managers and cannot be changed, with the exception of the Delete documents option, as shown in Figure 3.16. There usually should be only one user or group with Manager level access and one server with Manager level access, to prevent problems with security and replication. Servers with Manager access can replicate all changes to the database, including new and edited documents, design elements, and the ACL.

Figure 3.16   Manager access

 NOTE
A user can check access to a selected database by looking at the status bar. Click the icon to the left of the location to determine your access to the current database.

The Notes system administrator and the application developer should work together to create and maintain the ACL for each database. Some aspects of security can affect how the application functions, and this problem is under the jurisdiction of the application developer or database manager. On the other hand, some aspects of the ACL can affect the security of the Notes environment as a whole and can affect replication of the databases throughout the environment—both of which are the responsibility of the Notes system administrator.

VIEW AND FORM ACCESS LISTS

After a user is given access to a database through the ACL, the user then has to gain access to the forms and views in the database. This access can be granted or denied using View and Form access lists.

View and Form access lists are primarily the responsibility of the application developer or designer of the database. The system administrator should understand this layer of security, however, to ensure that replication occurs correctly and to ensure they understand the level of security necessary for the database.

To create form and view access lists, use the Security tab on the Form and View Properties InfoBoxes. These security tabs enable the designer to refine the access given to users.

In the view access list, the developer determines which users can use the view. All users with Reader access and higher to a database are given access to the views by default, as shown in Figure 3.17. To refine this access, remove the check mark from the All readers and above option. You then have the option to add check marks next to the users, servers, and groups that should be able to gain access to the view. The users must, however, be given access to the database through the database ACL to use the view.

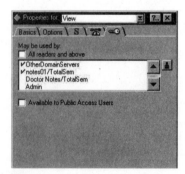

Figure 3.17   View access lists

View access lists should not be considered true security measures for a database. A user with at least Reader access to the database could create a private view or folder to display the data. The view and folder access lists should be used in combination with other security measures for the database.

The designer can refine access to forms using a similar method. To view the form access list, open the Form properties InfoBox to the Security tab, as shown in Figure 3.18. The first option on this tab is the default read access for documents created with this form.

All users with Reader access and higher are given this ability by default. You refine this access by removing this ability from all users with Reader access and giving it only to certain users. Note that a user must have access to the database through the database ACL to view documents in the database. If a read access list exists, however, a user must be on that list to read documents created by the form—regardless of ACL level. The second option on the Form properties Security tab is the list of the users who can create documents with the form. By default, all users with Author access and higher are given this ability.

Figure 3.18   Form access list

> **NOTE**
> Remember that this access list is only able to refine the ACL, not change it—which means that a user with Reader access in the ACL will not be able to create a document with the form, even if you have listed them in the form access create list. Similarly, if a user has Editor access or higher and is able to read the documents created with this form (i.e., the user is on the read access list), that user will be able to create documents with this form regardless of the create access list.

View and Form access lists not only must contain the users who should have access to the documents in the views or created by the forms but also must contain the names of the servers that will be replicating the databases.

READER AND AUTHOR FIELDS

Although default access to documents is granted through the settings set on the form and in the ACL, as described earlier, access in the database can be limited further through the use of readers and authors fields. Readers and authors fields can be either editable or computed and must resolve to a list of names. As in the view and form access lists described previously, readers and authors fields refine and limit the abilities given in the ACL; however, they cannot override the ACL. Reader and author fields are the responsibility of the application developer. The important things to remember when discussing them with the designers is that they cannot override the database ACL—and servers replicating the databases must be listed in these fields.

ROLES

Using and creating User Roles can further refine the security for an application. User Roles are created in the ACL dialog box and enable the designer to group users, servers, and groups that should have similar rights and abilities within a specific database. Roles provide the following advantages to the designer or manager of a database:

- Providing centralized security. Instead of changing user names in every form or view access list, in every readers or authors field, and in every section within a database, designers and database managers change the users within a particular role or change the access of the role.

- Roles are listed in the ACL of a database, ensuring that when changes are made to the security of the database, the changes take effect in the design elements. If the security were all done at the level of the design elements, many database managers and designers might neglect to update the security correctly.

To create a role, open the database Access Control dialog box to the Roles panel, as shown in Figure 3.19. To add a new role to the database, click Add . . . and type the name of the role you want to add. The name can be up to 15 characters. You can also rename or remove user roles from the database's ACL from this panel on the dialog box. The database manager can add up to 75 roles to the database.

After creating the user roles, you must apply those roles to the users, servers, and groups listed in the database ACL. To apply a role to a user, open the Basics panel on the ACL dialog box, as shown in Figure 3.20. Select the user and choose the role or roles that the user should have from the Roles list on the right-hand side of the dialog box. This option will associate this user role with the selected user, server, or group.

After creating and applying User Roles, the User Roles must be given functions within the database. User Roles can be used in the view and form access lists, in Readers or Authors fields, and in section access lists. The system administrator will create and apply User Roles in the access control list; however, the database designer will give the roles their function in the database. If you choose to implement this level of security, the system administrator and the database application developer should work together to plan the security.

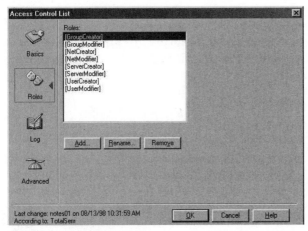

Figure 3.19 Roles InfoBox

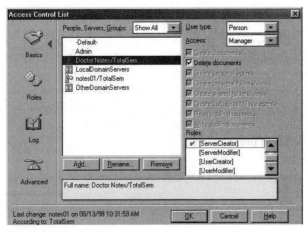

Figure 3.20 Assigning a user role

Securing the NAB

The NAB, as we mentioned earlier, is the most important database in your Notes environment. The NAB enables the system administrator to create users, servers, and groups, and modify and refine those objects and others. Lotus implemented additional security in the NAB to ensure that only the proper users, servers, and groups would have access to complete these tasks.

The NAB, like all databases, uses the ACL levels as its primary layer of security. There are also eight predefined user roles, however, that have been created and implemented for this database. These roles refine the rights given by the ACL by determining who can create and edit specific types of documents. When you assign the ACL for the NAB, be sure to include these roles for the appropriate users, servers, and groups:

- *GroupCreator.* The GroupCreator role must be added to a user with Author access or higher to give them the capacity to create new Groups in the NAB.

- *GroupModifier.* The GroupModifier role must be added for a user with Author access to give them the capacity to modify or delete Group documents in the NAB. Users with Author access must also be in the Owner's or Administrator's field for the Group documents they wish to edit.

- *NetCreator.* The NetCreator role must be added for a user with Author access or higher, to give them the ability to create any document in the NAB other than Person, Group, or Server documents —including Connection, Configuration, and Program documents, for example.

- *NetModifier.* The NetModifier role must be added for a user with Author access to give them the ability to modify or delete any document in the NAB, other than Person, Group, or Server documents. Users with Author access must also be in the Owner's or Administrator's field for the documents they wish to edit.

- *ServerCreator.* The ServerCreator role must be added for a user with Author access or higher, to give them the ability to create new Server documents in the NAB.

- *ServerModifier.* The GroupModifier role must be added for a user with Author access to give them the ability to modify or delete

Server documents in the NAB. Users with Author access must also be in the Owner's or Administrator's field for the Server documents they wish to edit.

- *UserCreator.* The UserCreator role must be added for a user with Author access or higher to give them the ability to create new Person documents in the NAB.

- *UserModifier.* TheUserModifier role must be added for a user with Author access, to give them the ability to modify or delete Person documents in the NAB. Users with Author access must also be in the Owner's or Administrator's field for the Person documents they wish to edit.

The roles for the NAB are shown in Figure 3.21.

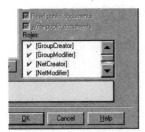

Figure 3.21   User Roles in the NAB

Signing and Encryption

Signing and Encryption are two additional security layers that can be applied to documents. For the system administration exam, you will concentrate on how these types of security are applied to mail. Understand, however, that these security layers can also be applied in other areas of Notes.

NOTE

Both signing and encryption rely on the keys contained in the ID files. As we described earlier, the key is a unique hexadecimal number. The three types of keys are Public, Private, and Secret. The Public and Private keys are mathematically related and are based on the dual-key encryption standard called RSA Cryptosystem technology. Secret keys are created by users and are stored in the ID file.

Signing

Signing is a process that places a verification stamp on a document or message, which enables the reader to confirm that the signer was the last person to modify the document. The signature is created using the user's Private key (discussed previously), in combination with the data in the field. The signatures can be attached either to sections, to fields within documents, or to an entire mail message. The following steps describe the process of signing a mail message:

1. The message is signed using the sender's Private key, which is contained only in the sender's user ID file. The signature that is appended to the message contains the signature itself, the user's Public key, and all of the sender's certificates.

2. The recipient then uses the certificates to verify the authenticity of the sender's Public key. The Public key is then used to verify the authenticity of the signature.

3. When the signature is verified, a message appears in the recipient's status bar to indicate the message has been signed according to the appropriate certifier.

4. When the signature cannot be verified, a message appears in the recipient's status bar to indicate that the message may have been modified or corrupted since it was signed.

To sign a mail message, a user can choose to enable the Sign option in the Delivery Options dialog box, shown in Figure 3.22. If a user wants all mail messages to be signed by default, they can enable the Sign Sent Mail option on the Mail panel in the User Preferences dialog box, shown in Figure 3.23.

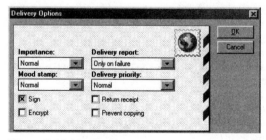

Figure 3.22 Sign a message

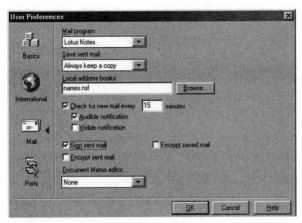

Figure 3.23   Sign all messages

In addition to signing mail messages, any field on any form can be enabled for signing. The database developer is responsible for enabling fields for signing.

Encryption

Encryption is a security level that can be applied to databases, fields, mail messages, and Notes network ports. *Encryption* means taking raw Notes data and encoding it, so that only a user with the correct key can decode and read the data. Some encryption uses a combination of the Public and Private keys for encryption and decryption, while other encryption in Notes uses a single, secret-key encryption method. Field-level encryption and network port encryption are done using a single, secret encryption key.

Field-level encryption enables an author or editor to apply one or more encryption keys to the document. These encryption keys are created using the Encryption panel on the User ID dialog box, as shown in Figure 3.24. When one or more encryption keys are applied, any data in encryption-enabled fields on the document is secured through encryption. All users who need access to data encrypted with this type of encryption key must have a copy of the encryption key in their user ID file.

Network port encryption uses a key that is generated for each new network session to encrypt all data that travels through a particular port. To enable Network port encryption, use the Encrypt network data option on the Ports panel of the User Preferences dialog box, as shown in Figure 3.25.

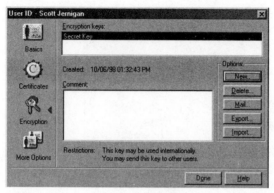

Figure 3.24   Encryption panel of the User ID dialog box

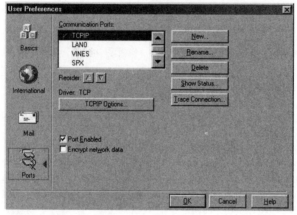

Figure 3.25   Encrypt network data

Other encryption, such as mail encryption, is done through a combination of the Public and Private keys stored in the Public NAB and in the user ID file. The Public key, which is created for each user when registered, is stored in the user's Person document in the Public Address Book and in the User's ID file. Public keys are available to all users who have access to the NAB. Part of a Public key is shown in Figure 3.26. The Private key is stored in the user ID file and is available only to the owner of that ID file. Either key can be used to encrypt or decrypt data. If data is encrypted using the Public key, however, only the Private key can decrypt the data.

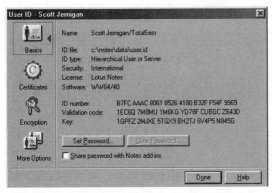

Figure 3.26 Part of a Public key

Mail encryption is used on documents that are being sent to another user's mail database. Mail encryption can be accomplished in one of three ways:

- *Encrypt incoming mail.* When a user chooses to encrypt incoming mail, the messages are encrypted at the server. The mail messages cannot be read by administrators or by the server. This option can be set in the user's Person document in the Public NAB, as shown in Figure 3.27.

FAX phone:	
Spouse:	
Children:	

Misc	
Comment:	
Encrypt incoming mail:	Yes
Other X.400 address:	
Calendar domain:	
Web page:	

Figure 3.27 Encrypt incoming mail

- *Encrypt outgoing mail.* When a user chooses to encrypt outgoing mail, the messages are encrypted as they are sent to other users. The message cannot be accessed by administrators or servers at any point in the transfer or delivery process. This option can be set in multiple ways:

  - File . . . Tools . . . User Preferences . . . Mail . . . Encrypt Sent Mail

- NOTES.INI variable SecureMail = 1
- Choosing Encrypt in the Delivery Options dialog box when sending mail messages, shown in Figure 3.28.

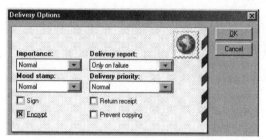

Figure 3.28 Encrypt a sent mail message.

- *Encrypt all saved mail.* Encrypting saved mail prevents messages stored in your server or local mail database from being accessed by unauthorized users. Many laptop users especially use this type of encryption to protect their data, in case of losing their laptop. To set this option, choose File . . . Tools . . . User Preferences . . . Mail . . . Encrypt Sent Mail. The encryption options available as a user preference are shown in Figure 3.29.

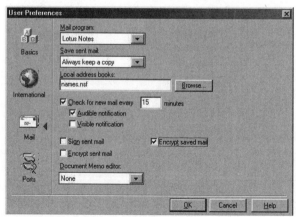

Figure 3.29 Encrypt all saved mail.

Signing, single-key encryption, and dual-key encryption are shown in Figure 3.30.

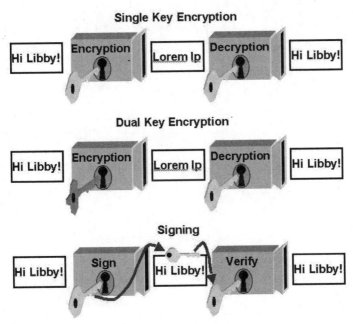

Figure 3.30 Security diagram

You also have the ability to encrypt an entire database, which is especially useful to laptop users who carry replicas of sensitive databases with them. To encrypt a local database, open the Database properties InfoBox and click the Encryption button. The Encryption dialog box shown in Figure 3.31 is displayed. Choose to Locally encrypt the database and select a level of local encryption. The three levels are strong, medium, and simple. Strong and Medium encryption cannot be used on systems that use disk compression. Simple encryption provides the fastest access to your data. Only the user with the appropriate ID file selected on this dialog box will be able to read the database. This type of encryption is created using the Public and Private keys stored in the ID file.

NOTE
Note that only users with Manager access to a database can encrypt the database.

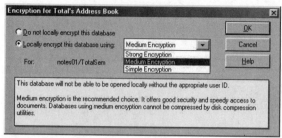

Figure 3.31   Encryption dialog box

Review Questions

1. List the components of an ID file.

2. The Public key is stored in a user's ID file, and only they have access to it.

    a. True

    b. False

3. A user can be switched from International to North American License type at any time.

    a. True

    b. False

4. A user's password is stored both in the ID file and in the NAB.

    a. True

    b. False

5. Notes protects passwords from password-guessing programs using which feature?

    a. Multiple password protection

    b. Case-sensitive password

    c. Time-delay

    d. Anti-spoofing

6. Notes protects passwords from password-capturing programs using which feature?

    a. Multiple password protection

    b. Case-sensitive password

    c. Time-delay

    d. Anti-spoofing

7. Authentication will be successful in which situations?

   a. When the user and server are created by the same OU certifier

   b. When the user and server are created by the same O certifier

   c. When the server is registered by Houston/TotalSem, and the user is registered by Accounting/Dallas/TotalSem

   d. When the server is registered by Houston/ACME, and the user is registered by Houston/TotalSem

8. Authentication is a one-way process from the user to the server.

   a. True

   b. False

9. Aaron has run out of room on the drive on which his \NOTES\ DATA directory exists. What can you suggest?

   a. Create a new \NOTES\DATA directory on a larger drive. Move all the data to the new drive.

   b. Create a new \NOTES\DATA directory on a larger drive. Create a directory link to this drive.

   c. Create a directory called anything but \NOTES\DATA on a larger drive. Create a directory link to this drive.

   d. Reinstall Notes on a larger drive.

10. Cindy tries to open a database listed in the Open Database dialog box called Accounting. She is denied access. When she asks the help desk to check, it shows that she is in the ACL of the database as an Editor. When the help desk calls you, what do you tell them to check?

    a. Tell them they looked at the wrong user in the ACL.

    b. Tell them to look for a database link file.

    c. Tell them to look for a directory link file.

    d. Tell them to look in the Server Access list.

11. For greater security in your Notes databases, you should always rename the Default group in each database.

    a. True

    b. False

12. Sylvia is the manager of the Notes environment and allows three other Database managers to help her with creating ACLs for databases. She notices that one database uses individual user names in the ACL, instead of group names. Can she find out which database manager created the ACL for this database?

    a. Yes. She can look in the ACL dialog box Basics panel for Log information.

    b. Yes. She can look in the ACL dialog box Log panel for Log information.

    c. Yes. She can look in the Database properties InfoBox Log panel for Log information.

    d. No.

13. What is the lowest level of access a user needs to be able to edit the documents they create in a database?

    a. Editor

    b. Author

    c. Designer

    d. Manager

14. What is the lowest level of access a server needs to be able to replicate new documents?

    a. Editor

    b. Author

    c. Designer

    d. Manager

15. As a system administrator, what is important to you about view and form access lists?

16. Dudley looks at the ACL for the NAB and sees that he has Editor access. When he tries to create a new Group in the NAB, however, he cannot. Why not?

    a. The NAB has additional security. To create a new group, he has to have Manager access.

    b. The NAB has additional security. To create a new group, he has to be in the Administrator's field of the server document.

    c. The NAB has additional security. To create a new group, he needs the GroupCreator role.

    d. User error.

17. The NetModifier role enables a user with Editor access to the NAB to modify the Network configuration section of the Server document.

    a. True

    b. False

18. When Robyn receives a mail message signed by a user from another organization and cannot verify the signature, what happens?

    a. She can open the document but not read it.

    b. She cannot open the document.

    c. She can open the document, but receives an error message that forces her to cross-certify with the sender.

    d. She can open the document but receives an error message that indicates someone may have tampered with the message.

19. Mail messages can be signed or encrypted by default.

    a. True

    b. False

20. If a document is encrypted using the user's Public key, which key is used to decrypt the data?

    a. Public

    b. Private

21. How many users have access to the user's Private key?

    a. One

    b. All

22. Network encryption uses what kind of encryption?

    a. Dual key

    b. Single key

23. What level of access do you have to have to locally encrypt a database?

    a. Editor

    b. Manager

    c. Designer

    d. Author

Review Answers

1. User name, North American or International license, Public and Private keys, any additional secret encryption keys, original hierarchical certificate, password

2. b. False. The Public key is stored in both the user's ID file and the NAB.

3. b. False

4. b. False

5. c. Time-delay

6. d. Anti-spoofing

7. a, b, and c. In each of these cases, the user and server have a certificate in common. In d, they will have to cross-certify (Chapter 7, "Advanced Configuration and Setup") before they can authenticate.

8. b. False. Authentication is a two-way process that challenges both the user and server.

9. c is the best answer. Create a directory called anything but \NOTES\DATA on a larger drive. Create a directory link to this drive.

10. b or c. Tell them to look for a database link file or tell them to look for a directory link file. Either file can contain names that enable access only to those users, regardless of ACL.

11. b. False. The Default group cannot be renamed.

12. b. Yes. She can look in the ACL dialog box Log panel for Log information.

13. b. Author

14. b. Author

15. The servers that will be replicating the database must be in any secured view and form access lists.

16. c. The NAB has additional security. To create a new group, he needs the GroupCreator role.

17. b. False. The NetModifier role enables a user with Author access to modify documents other than Person, Group, and Server documents in the NAB.

18. d. She can open the document but receives an error message that indicates someone may have tampered with the message.

19. a. True
20. b. Private
21. a. One
22. b. Single key
23. b. Manager

CHAPTER 4

Notes Mail

Objectives

After reading this chapter, you should be able to answer questions based on the following objectives:

- Enable and configure Message-based mail.
- Enable and configure Shared mail.
- Use server console commands to force mail routing.
- Schedule mail routing.
- Convert from Message-based mail to Pointer-based Shared) mail.
- Set up scheduled routing using Connection documents.
- Route mail automatically via Notes Named Networks.
- Maintain Message-based mail.
- Maintain Shared mail.
- Use server console commands to configure and maintain Shared mail.
- Troubleshoot mail problems.

Introduction to Notes Mail

The System Administration I exam covers the topic of mail in significant detail. Mail is the first, most vital process to most companies and most users—which is one of the reasons for covering it in so much depth on the exam. Mail is also covered in depth because it encompasses numerous processes and tasks that an administrator must be able to perform. The mail objectives on the exam come from two perspectives: Message-based mail and Shared mail. These terms refer to the two different storage possibilities for mail in Notes R4.X. Because Shared mail is new to R4.X, it is vital for you to understand its configuration and troubleshooting, both independently and in comparison with Message-based mail.

This chapter covers six basic mail topics. First, you need to understand the basic terminology Notes uses for mail. Second, we will discuss mail routing. Notes mail routing is determined by two types of server groupings: domains and *Notes Named Networks* (NNNs). System Administration I covers mail within the same domain, both within and between NNNs. The System Administration II exam will cover mail routing between and among different domains. For the exam, it is vital to understand how each task related to mail routing is performed and which Notes component performs the task. You must also know the documents that determine how and when mail is routed. To route mail outside the NNN or among domains, you need a mail routing topology. We will cover the main routing topologies later. Finally, we will cover the topics of Shared mail and mail troubleshooting.

Mail Terminology

To administer a Notes environment properly—and pass the System Administration I exam—you must understand the mail terminology used in Notes. The following chart (Table 4.1) outlines the terms you need to know.

Table 4.1 Mail Terminology

Term	Definition
Notes Named Networks (NNNs)	Servers that are in constant contact (i.e., same LAN or WAN) and share the same protocol can be in the same NNN. When servers are in the same NNN, mail is routed immediately and automatically, regardless of Connection documents. We covered how to create NNNs in Chapter 1, "Installation and Configuration." In terms of mail, however, the main point is how mail is routed in a NNN—immediately and automatically.
Message-based mail	Message-based mail is the most common form of mail in older mail products and is probably what you think of when you think of mail. In Message-based mail, the entire mail message sent to a user —including both the body and the headers—is stored in the user's mail file.
Shared mail (Pointer-mail, single object copy store)	In Notes R4.X, Lotus introduced Shared-based mail. In Shared mail, rather than storing the entire mail message sent to each user in that user's mail file, only the header of a message is stored in the user's mail file. The body of the message, including attachments, is stored in the SCOS, or Shared mail database. This can be a great space-saver on servers, especially when users tend to send a single message to multiple recipients.
Domains	All servers that share the same Public NAB are in the same Notes domain. Often, an entire organization will stay within a single domain, although later in this book we will discuss situations in which this is not the case. In terms of mail, addressing of mail messages within a domain is based on that single NAB, which simplifies addressing and type-ahead.[1]

continues

Table 4.1   Continued.

Term	Definition
Router	The router is the server-based task that delivers and transfers mail. The router also searches the Public NAB for recipients' home servers, as well as connections to other servers or domains when necessary. Please note: If you are familiar with hardware and TCP/IP or NT and other operating systems, the term *router* may have a different meaning for you. In this book, the term router will always mean the server task used for mail delivery and transfer, unless otherwise specified.
Mailer	The mailer is the workstation-based component that deposits a mail message on the sender's server's MAIL.BOX. The mailer touches the messages first after the user composes the message. In addition, the mailer is responsible for looking up users in the Public NAB to verify existence and spelling, as long as the users are in the same domain.
MAIL.BOX	The MAIL.BOX is a server database that temporarily holds messages as they are in the process of being delivered and transferred. This database is often referred to as the server's outgoing mailbox.
Mail delivery	Mail delivery is the process of placing a message in the recipient's mail file. Delivery is accomplished by the router when mail is moved from the MAIL.BOX on the recipient's server to the recipient's mail file.
Mail transfer	Mail transfer is the process of moving mail between servers, from MAIL.BOX to MAIL.BOX. The Router takes messages bound for recipients on other servers from the MAIL.BOX of the sender's server and places it in the MAIL.BOX of the recipient's server (and any intermediate servers).

[1]Those of you who are familiar with TCP/IP or NT and other operating systems terminology, please note that domains in this book will always refer to Notes domains unless otherwise specified.

Mail Routing

Before we discuss the server processes related to mail, it may be useful to refresh your memory about the front-end, or client, processes. These processes are not tested directly on the System Administration I exam. The exams assume, however, that you know how to use the client processes in Lotus Notes.

To begin a mail memo, the user opens the mail file and chooses the New Memo action button, which displays the blank message form. The user either chooses the address button to place recipients in the TO: field, or begins typing in the username, waiting for the Notes type-ahead feature to complete the name. The user then types the body of the memo. Before sending, the user may choose the Delivery Options action button and request High or Low-priority or a return receipt. Other choices include the security options of signing, encrypting, and preventing copying. The user either saves, saves and sends, sends, or discards the message.

During this mail process, the mailer verifies the names in the TO:, CC:, or BCC: fields and then places the message in the server's mail routing database, the MAIL.BOX. After the message has been delivered to the MAIL.BOX database, the router determines where it goes and how it arrives. The router decides where and how to route mail, taking into account the following items:

- Is the mail going to a recipient on the same server (and thus in the same NNN)?

- Is the mail going to a recipient on a different server in the same NNN?

- Is the mail going to a recipient in another NNN, but still in the same domain?

- Is the mail going to a recipient in another domain altogether?

Mail Routing Within the Same NNN

 NOTE
The following processes refer to Message-based mail. While the basic process remains the same in Shared mail, the placement of the mail messages is different. We will discuss these differences later in this section.

The mail routing process begins when the user composes a mail message in the mail database. Once the mail message is created and addressed, the mailer checks the recipients' names and addresses and places the message in the sender's server's MAIL.BOX. Once the message is in the MAIL.BOX, the router takes over and determines the location of the recipient's mail file (whether on the same server or on a different server). This part of the process is the same in all server-based mail scenarios. The rest of the process is then defined by the location of the message recipient.

MAIL ROUTING ON THE SAME SERVER

When the router determines that the recipient's mail file is on the same server, the message is immediately delivered to the recipient's mail file. The router determines the location of the recipient's mail file from their Person document in the Public NAB, as shown in Figure 4.1.

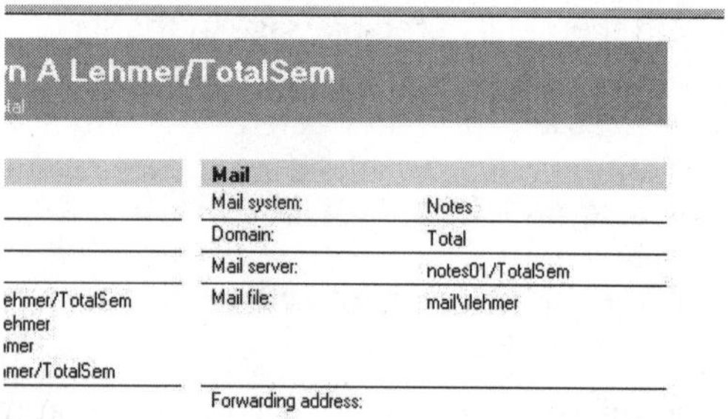

	Mail	
	Mail system:	Notes
	Domain:	Total
	Mail server:	notes01/TotalSem
ehmer/TotalSem	Mail file:	mail\rlehmer
ehmer		
rmer		
mer/TotalSem		
	Forwarding address:	

Figure 4.1 Person document showing home server

In a company called Total Seminars, for example, there is only one mail server: ServerA/TotalSem. If Cindy Smith/TotalSem sends Paul Jones/TotalSem a message in this environment, the mailer on her workstation will first verify that Paul Jones/TotalSem is in the Public NAB and that his name is spelled correctly. Then the mailer will deposit the message in the MAIL.BOX on ServerA/TotalSem. The router task will then check to see whether ServerA/TotalSem is Paul Jones' home server. When this item is verified, the router will deliver the message to Paul Jones' mail file. Figure 4.2 shows a diagram of this example.

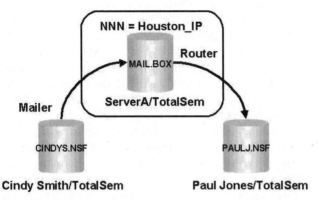

Cindy Smith/TotalSem   **Paul Jones/TotalSem**

Figure 4.2   Mail routing on the same server

MAIL ROUTING BETWEEN DIFFERENT SERVERS

When the router determines from the recipient's Person document that the recipient resides on a different home server than the sender, the message is transferred from the MAIL.BOX on the sender's server to the MAIL.BOX on the recipient's server. The router uses the server documents and Connection documents from the Public NAB to determine a path to the other server. In the case where both servers have the same NNN, the router is able to transfer the mail to the remote MAIL.BOX automatically and immediately, without the need for a Connection document. The server can determine the NNN from the Network section of the Server document, as shown in Figure 4.3. After the message is in the recipient's home server's MAIL.BOX, the router task for that server delivers the message to the recipient's mail file.

▶ Server Location Information
▼ Network Configuration

Port	Notes Network	Net Addres
TCPIP	TotalIP	192.168.42.:
		notes01
		notes01
		notes01
		notes01
		notes01

Figure 4.3   Server document showing NNN

When Total Seminars expands, for example, it builds a second mail server. This server—ServerB/TotalSem—is on the same LAN as Server A and uses the same protocol, TCP/IP. The Notes administrator, therefore, places ServerB in the same NNN, Houston_IP.

When Cindy Smith/TotalSem sends a message to Chloe Burns/ TotalSem, whose home server is ServerB/TotalSem, the message is created in Cindy's mail file. The mailer verifies the spelling of Chloe's name and places the message in ServerA/TotalSem's MAIL.BOX. The router determines that Chloe's home server is ServerB/TotalSem by looking at her Person document in the Public NAB. Next, the router looks for a way to get to ServerB/TotalSem. From the server documents in the NAB, the router determines that ServerA and ServerB are both in the Houston_IP NNN. The router can then immediately transfer the message to ServerB's MAIL.BOX. ServerB's router task then verifies that Chloe's mail file is on that server and delivers the message to her mail file. A diagram of this example is shown in Figure 4.4.

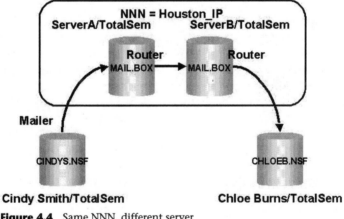

Figure 4.4 Same NNN, different server

Mail Routing Between Different NNNs

When servers are not constantly connected—and/or when they use different protocols, they must be placed in different NNNs, which changes the mail routing process. The process changes when the router determines from the server documents that the home servers of the sender and the recipient are in different NNNs. At this point, the router needs to find a path to the remote server. This path is found in a Server Connection document, an example of which is shown in Figure 4.5.

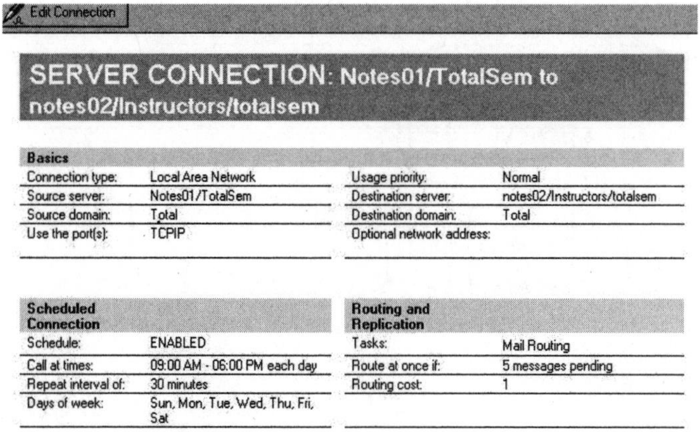

Edit Connection

SERVER CONNECTION: Notes01/TotalSem to notes02/Instructors/totalsem

Basics			
Connection type:	Local Area Network	Usage priority:	Normal
Source server:	Notes01/TotalSem	Destination server:	notes02/Instructors/totalsem
Source domain:	Total	Destination domain:	Total
Use the port(s):	TCPIP	Optional network address:	

Scheduled Connection		Routing and Replication	
Schedule:	ENABLED	Tasks:	Mail Routing
Call at times:	09:00 AM - 06:00 PM each day	Route at once if:	5 messages pending
Repeat interval of:	30 minutes	Routing cost:	1
Days of week:	Sun, Mon, Tue, Wed, Thu, Fri, Sat		

Figure 4.5 Server Mail routing Connection document

When servers are in different NNNs, mail routing requires a Server Connection document for each direction of mail. To each server, mail routing is a one-way process. You send a message, and it stops at its destination. When a reply is sent, it is another one-way process started by the other server. Because mail is one-way, all mail routing will require *two* Connection documents (one for each server). The Connection document also determines the schedule on which mail is transferred between the two servers. After the router determines the path to the recipient's home server and the mail routing schedule to follow, it transfers the mail message to the recipient's home server's `MAIL.BOX`. For example, when Total Seminars opens a second office in a different building, it connects the new server, ServerC, to the servers at the first location using modems. Because the servers are not constantly connected, the Notes Administrator places them in separate NNNs and creates Connection documents for mail routing.

Cindy Smith/TotalSem (whose home server is ServerA/TotalSem) sends a mail message to Oscar Johnson/TotalSem (whose home server is ServerC/TotalSem). The router determines that Oscar is on a different mail server and that the mail server is in a different

NNN. The router then uses the Connection document from ServerA to ServerC to find a path to ServerC. After finding the path, the router waits until the next scheduled mail routing time and transfers the mail to ServerC/TotalSem's MAIL.BOX. ServerC's router then places the message in Oscar's mail file. A diagram of this example is shown in Figure 4.6.

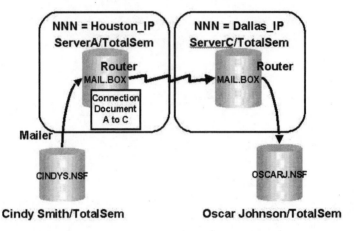

Figure 4.6 Different NNN, different server

Another related scenario is that another user in the environment, on ServerB, tries to send a message to a user on ServerC. There is no Connection document connecting ServerB and ServerA. ServerB can, however, use the Connection document for another server in its NNN to route mail outside the NNN. In this case, when Paul, whose home server is ServerB, tries to send a message, the router uses ServerA's Connection document to find a path to ServerC, as shown in Figure 4.7.

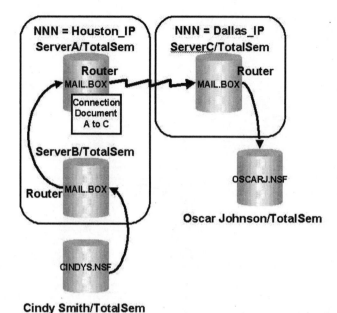

Cindy Smith/TotalSem

Figure 4.7 Different NNN, different servers, and ServerB using ServerA's
Connection document

Connecting to Different Servers

As described in our examples earlier and in the requirements to
route to servers outside your NNN, your servers must have a way to
connect. This path is determined by the Connection documents
created in the Server's Public NAB. Let's take a moment to examine
the server Connection document. Connection documents for mail
are required any time the servers are not in the same NNN. Mail
Connection documents define the path to the other server, the
routing schedule, and any mail thresholds. As mentioned earlier, to
create a complete mail route that enables users to send and receive,
two mail Connection documents must be created for each connected server. To create a Connection document, go to the Public
NAB. In the Navigation pane (on the left), expand the Server section. Choose the Connections view. Once in the Connections view,
click the New Connection action button to create a new server
Connection document, as shown in Figure 4.8.

SERVER CONNECTION

Basics

Connection type:	Local Area Network	Usage priority:	Normal
Source server:	Notes01/TotalSem	Destination server:	
Source domain:		Destination domain:	
Use the port(s):		Optional network address:	

Choose ports

Scheduled Connection		Routing and Replication	
Schedule:	ENABLED	Tasks:	Replication, Mail Routing
Call at times:	08:00 AM - 10:00 PM each day	Route at once if:	5 messages pending
Repeat interval of:	360 minutes	Routing cost:	1
Days of week:	Sun, Mon, Tue, Wed, Thu, Fri, Sat	Replicate databases of:	Low & Medium & High priority
		Replication Type:	Pull Push
		Files/Directories to Replicate:	(all if none specified)
		Replication Time Limit:	minutes

Figure 4.8 Server Connection document

Table 4.2 describes the fields on the Connection documents that relate to mail routing. You will notice that not all fields are described. The only fields described are those that relate to mail routing and also are tested on the exam.

NOTE

Here is a *gotcha* to be wary of when creating Connection documents. One of the end user mail options when sending mail is for Delivery Priority. The priority choices are Low, Normal, and High. By default, all mail is sent Normal priority. Normal-priority mail follows the mail settings in the Connection document—including schedule and mail threshold. High-priority mail is routed immediately, causing the server to make an immediate connection to the remote server and send all Normal—and High-priority pending mail. Low-priority mail is routed by default from midnight to 6 a.m. If there is no mail Connection document that covers the low-priority times, low-priority mail will not route.

EXAM TIP

When you route mail through other servers to get to a final recipient, be certain that you limit the router hops to fewer than 25. Twenty-five hops is the hard-coded limit of mail routing hops, and messages will fail once this limit is reached. The hops are counted down from 25 each time the message passes through a server's MAIL.BOX.

Table 4.2 Server Connection Document Fields

Field Name	What You Should Put Here	Notes
Connection Type	This field name describes the manner in which the two servers are connected: for example, LAN, dialup modem, and Passthru server. Choose the appropriate type for your servers. Please note that the value in this field changes the other fields.	You can have multiple Connection documents between any two servers, using different connection types in case of WAN outages, etc.
Source Server	This field name should be the hierarchical name of the server that is originating the connection. Because mail is one-way, routing mail between two servers requires a Connection document for each server, and each server will be a source server for one of the connection documents.	The source server is the server on which you are creating the document. For example, if you have the NAB open on Server A when you create the Connection document, Server A automatically will be the source server. You can (and often must) change this field.
Source Domain	The source domain is the Notes domain in which the source server exists.	The source domain is empty by default.

continues

Table 4.2 Continued.

Field Name	What You Should Put Here	Notes
Use the Ports:	Define the port that this connection will use. In the case of a LAN Connection document, this relates to the protocol—TCP/IP or IPX/SPX, for example. For a dialup modem connection type, this is usually COM1.	
Usage Priority	You can set a low usage priority if you want a particular connection to be used only as a last resort. For example, if two servers are connected over a bridged/routed WAN, this would probably be your first choice of methods to connect for mail routing. Modems also exist on both servers, however, as a backup scenario in case of the WAN link being down. You would make the dialup modem Connection document low-priority in this case.	Normal or Low. Default is Normal.
Destination Server	The hierarchical name of the server to which you are trying to route mail.	
Destination Domain	The Notes domain in which the destination server exists.	

Field Name	What You Should Put Here	Notes
Schedule	Choose whether the schedule is enabled or disabled. At times you may need to disable a certain connection, such as when a server is off-line, but you may not want to delete the Connection document.	
Call at times	These are times at which the server should make contact with the destination server. This can be specific times or ranges of time.	
Repeat Interval	The value in this field describes how often the source server will attempt to contact the destination server. If only one time or one call is desired, this field can be left blank.	
Days of Week	Choose the days of the week that this Connection document is valid. You may choose to have different connection times for the weekends, for example.	
Tasks	Your choices in this field include Mail Routing, Replication, X.400 Mail Routing, SMTP Mail Routing, and cc:Mail Routing.	Choose Mail Routing in this case. The default is both Replication and Mail Routing.

continues

Table 4.2 Continued.

Field Name	What You Should Put Here	Notes
Route at Once If	This field enables you to specify a maximum number of messages to hold in the queue in the MAIL.BOX before routing is forced. Mail will route immediately when this number is exceeded.[1]	The default value is five.

Field Name	What You Should Put Here	Notes
Routing Cost	The routing cost is one of the determining factors for which Connection document the router chooses when trying to get to another server. The router will choose the lowest routing cost path first.	Default for LAN = 1 Default for Dialup = 5

[1]Note that low-priority mail *does not* route when the threshold is reached; however, it does add to the count for the threshold. For example, if there are 4 low-priority messages waiting and you send a normal priority message, the normal message will route immediately, because the default threshold will have been reached. The four low-priority messages will remain in the queue.

Mail Routing Between Domains

When mail servers are in different domains, the addressing and delivery becomes more complex. This issue will be discussed more in depth for the System Administration II exam. A brief discussion here, however, may make it easier to visualize mail in general. When you see a Notes mail address that has been resolved, or checked, by the mailer, it looks something like this: Libby Schwarz/ TotalSem@Total. The last part of the e-mail address, after the @ symbol, is the Notes domain.

NOTE
Do not get confused thinking about domains in the NT terminology or the DNS terminology. They are not directly related, although the names may be the same.

When mail routes within the same domain, the domain portion of the address serves only to verify the domain for the router. This portion of the address is vital, however, if you intend to send messages outside your Notes domain. In most cases, sending mail outside your own domain will also mean sending mail outside your company. This action may entail sending mail via the SMTP *message transfer agent* (MTA), or you may be connected via Notes Net or other native Notes methods. You must have a Connection document describing a path and schedule to the other domain to be able to route mail to the other domain. In this respect, it is the same process as routing mail to a server on a different NNN. In addition, however, the router needs a Domain document, which shows the router the existence and definition of this other domain. Sending mail outside your domain will be covered in Chapter 7, "Advanced Configuration and Setup," for the System Administration II exam.

Mail Topology

First, what do we mean when we discuss topology? Topology in Notes, as in other computing realms, relates to where servers are located and how they are connected. We will discuss topology both for Notes mail routing and for Notes replication. These topologies may be the same in your Notes enterprise, or they may be different.

The basic reason for a mail topology is so you can control and predict how mail will be routed within your organization, especially when you have multiple NNNs or multiple domains. In many organizations, there are bandwidths and other communications reasons for controlling the topology as well. The main types of mail routing topologies are hub and spoke, end-to-end, ring, and mesh. Expect to see a diagram of one or more of these topologies on the System Administration I exam.

Hub and spoke is probably the most common topology in enterprise environments because it is the most efficient and organized. In hub and spoke, one server (the hub) schedules, initiates, and controls all mail routing. For example, you may have a corporate office in Houston and smaller offices all over the world. You might choose the hub server in Houston as the central hub for your enterprise. In this case, the hub will call the other servers (or initiate the connection over a LAN or WAN), and initiate mail routing. All the other servers will be the spokes. They will forward mail only through (to and from) the hub when they are called by the hub. In addition, the spokes will only route mail to one another via the hub—never directly. The hub and spoke topology is shown in Figure 4.9. The hub and spoke can be expanded to have multiple levels, in which case it is called a binary tree topology.

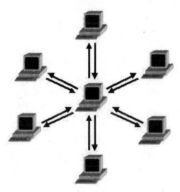

Figure 4.9 Hub and spoke routing topology

End-to-end routing topology has each server connected in a line, as shown in Figure 4.10. Each server will talk to the server ahead of it and after it in the line. In this chain, for example, the server in our Houston office would call the server in our Dallas office. The Dallas server would call the Kansas City office. The chain would then reverse on itself, and Kansas would call Dallas, and so on.

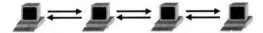

Figure 4.10   End-to-end routing topology

A ring topology is very similar to the end-to-end topology, as shown in Figure 4.11. When the last server is reached in the line, however, it simply circles back to the beginning. So in our example, after Dallas calls Kansas City, Kansas City would turn around and call Houston.

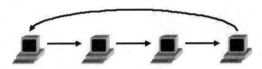

Figure 4.11   Ring routing topology

In Mesh topology, all servers connect directly to all other servers —which can be a difficult topology to administer due to its complex and disorganized nature. Mesh topology is not recommended but is shown in Figure 4.12.

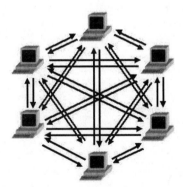

Figure 4.12   Mesh routing topology

You will organize your mail routing Connection documents using one of these routing topologies or a combination of routing topologies, which will ensure that all the users in your organization can send and receive mail in a timely fashion.

Shared Mail

Shared mail is the single most tested topic on the System Administration I exam, and it is completely new to Notes R4.X. To pass the
exam, it is important to understand the concept of Shared mail, the
various settings and terminology, and how to enable and maintain
Shared mail.

Shared mail is sometimes also called Pointer-based mail or object-
store mail. In Message-based mail, discussed previously, each mail
message is stored in its entirety in the user's mail file. In Shared
mail, messages are split between the user's mail file and another
database called the *Single Copy Object Store* (SCOS).

The main purpose of Shared mail is to save space on the mail
server. When Message-based mail is used, for example, any messages
sent to multiple recipients get stored multiple times on the server. If
I send a broadcast message to 15 users, that message is stored 15
times. In Shared mail, however, the message itself would be stored
only once. Each user would only have a pointer to the message, contained in a header. The body of the message is stored once, in the
SCOS. If we assume a simple text message of 10K, with Message-
based mail, the storage space required on the mail server is 150K. In
Shared mail, there would only be 10K for the message itself and
another 15K total for the pointers in each user's mail file, for an
overall savings of 115K. Then imagine the savings if the message
included rich text with formatting or attachments, which can raise
the size of the messages exponentially, as you can see in Figure 4.13.

 NOTE
The header, or pointer, portion of the message contains the following fields: TO:, CC:, BCC:, SUBJECT, and FROM. In addition, it
contains a link, or pointer, to the content of the message, which
is stored in the Shared mail database. The content is the message body,
rich text components, and any attachments.

To the reader of the message, Shared mail is transparent. There is
no difference in the way a user sees or reads messages when Shared
mail is enabled.

One item to consider when deciding whether Shared mail is a
useful option is the pattern of use for mail in your environment. If
users send messages to multiple recipients who have the same
home server, you will have the greatest savings. If your users send
many mail messages to only one or two recipients, you will still
have some savings of space—but not as much as in other scenarios.

In addition, if your users are split among many mail servers and often send messages to users on those other servers, you will have a smaller amount of space savings. As we examine the Shared mail configuration settings and processes, you will find other situations where Shared mail would be more or less useful. These scenarios represent typical questions on the exam.

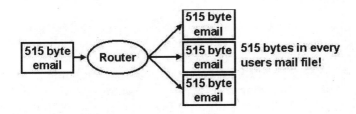

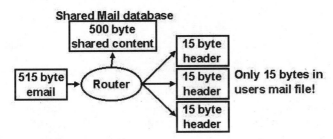

Figure 4.13   Shared mail savings

Configuring Shared Mail

Shared mail can be enabled in multiple ways. The easiest, and most complete in terms of configuration, is by typing the following command at the server console:

```
TELL ROUTER USE SHAREDMAIL.NSF
```

Insert the name of your Shared mail database in place of SHAREDMAIL. NSF in the commands. The Shared mail database can have any name other than MAILOBJ.NSF. This command creates the requisite files for Shared mail and enables Shared mail for all new messages. Other ways of enabling Shared mail will be discussed at the end of the section.

When Shared mail is enabled on a server, two files are created. These two files are MAILOBJ.NSF and SHAREDMAIL.NSF. By default,

SHAREDMAIL.NSF is named MAILOBJ1.NSF, and it can have any name—
with the exception of MAILOBJ.NSF.

> **EXAM TIP**
> The System Administration I exam often uses the default name
> for this file, but it may also use a name such as SHAREDMAIL.NSF.
> This similarity can cause confusion when answering questions
> regarding Shared mail.

The file MAILOBJ.NSF is a pointer file. While it has a .NSF extension,
it cannot be opened in Notes. Rather, it is a text file and can be
opened either by using the Open With . . . option in Windows NT
and 95 (available in Explorer or My Computer by holding down
Shift while right-clicking the file), or by opening NOTEPAD.EXE and
then using File . . . Open to open the file. When you open this file,
you will find a path to the other file, SHAREDMAIL.NSF. Whenever a
new Shared mail database is enabled, this path must change to
reflect the active Shared mail database.

The second file that is created by enabling Shared mail is SHAREDMAIL.
NSF, again, by default MAILOBJ1.NSF. You can choose to name this file
anything that makes sense in your environment, but be aware of two
concerns. First, the exam most often uses the default name, which
will increment by one if you choose to create another Shared mail
database. Second, the filename that you choose cannot be MAILOBJ.NSF
and must have an NSF extension. This database—the object store, or
Single Copy Object Store (SCOS) database —will be the repository for all
the Shared mail messages. As you might imagine, it can grow to be
large and requires careful backup and preventive maintenance.

There are three main configuration options for Shared mail. In
the NOTES.INI file, Shared mail can be set to zero, one, or two. When
the SHARED_MAIL setting is set equal to zero, Shared mail is disabled
and Message-based mail is used. This setting is the default setting in
Notes R4.X until Shared mail is enabled.

With the setting SHARED_MAIL=1, the server console reports that
Shared mail will be used for mail transfer. This report indicates two
things. First, the Shared mail database will be used for messages
intended for two or more recipients only. If the mail message is
only intended for one recipient, it is written in full to the user's
mail file, as shown in Figure 4.14. If the message is intended for two
or more recipients, the message is written to the first user's mail file
until the router discovers the second recipient. When the router
discovers the second recipient, it writes the message to the Shared

mail file and replaces the message in the first recipient's mail file with the header information only, as shown in Figure 4.15. The header information is then written to other recipients' mail files.

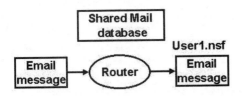

Figure 4.14 Mail recipients=1, SHARED_MAIL=1

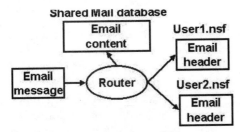

Figure 4.15 Mail recipients>1, SHARED_MAIL=1

Setting SHARED_MAIL=1 also determines that Shared mail will only be used when recipients are on the local server as their home server. If the message has to be transferred to another server, it will reside in the MAIL.BOX until the router makes the transfer. Similarly, if the server is an intermediate step in reaching a recipient's final destination, the Shared mail database will not be used.

By default, when you enable Shared mail using the *tell router use* command, SHARED_MAIL will be set equal to two. When SHARED_MAIL=2, the console reports that Shared mail is enabled for delivery and transfer. Again, this report indicates multiple things to the server. First, all mail messages, regardless of the number of recipients, will use the Shared mail database. Therefore, messages that are created for just one user will use the Shared mail database in the same way that messages for two or more recipients will, as shown in Figure 4.16.

In addition, because Shared mail is enabled for both delivery *and* transfer, messages that are intended for recipients on other mail servers will use the Shared mail database until the router transfers the message. Messages stopping on the server also will use the Shared mail database as an intermediate step to another server. In many environments, this result will not be desirable, if the server is often an intermediate step to other servers.

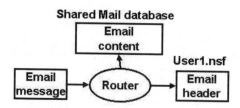

Figure 4.16 Any number of mail recipients, SHARED_MAIL=2

Creating a server configuration document that changes the NOTES.INI SHARED_MAIL setting can also enable Shared mail. To create a configuration document, open the Public NAB and expand the Servers view twistie on the navigation pane. Choose the Servers . . . Configurations view and click the Add Configuration action button. This action creates a blank configuration document. Choose the appropriate server (or leave the asterisk for all servers maintained by this NAB), and click the Set/Modify Parameters button. Under the Item keyword field, choose SHARED_MAIL, and set the value to one or two, depending on the type of Shared mail configuration appropriate for your environment. Save the configuration document, as shown in Figure 4.17.

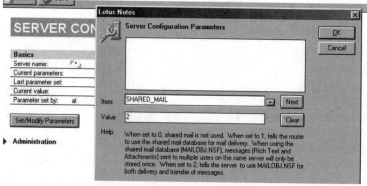

Figure 4.17 SHARED_MAIL settings in a configuration document

NOTE
Be aware that the server will not implement the configuration document immediately. The configuration change will not be implemented until you restart your server if you use the configuration document.

You also have the option of changing or adding the SHARED_MAIL configuration directly in the NOTES.INI file itself. This leaves more opportunities for mistakes and making errors in the server configuration, which can affect other aspects of the server.

By default, when you enable Shared mail, all new mail messages sent on the server will use Shared mail. You do not have to type any commands or do anything to users' mail files. On the other hand, if mail has been routed on your mail server before Shared mail was enabled, you must link these messages to the Shared mail database for those messages to be stored in the Shared mail file. The console commands used to link these messages to the Shared mail file are discussed later in this chapter.

Mail Routing with Shared Mail

Mail routing when Shared mail is enabled is similar to Message-based mail routing. When the user sends the message, the mailer makes the first check of the names in the NAB and transfers the message to the MAIL.BOX. Table 4.3 outlines the next steps taken by the router.

When SHARED_MAIL=1, messages that are being routed to other servers remain in the MAIL.BOX until routing takes place. When SHARED_MAIL=2, messages that are being routed to recipients on other servers are stored in the Shared mail database until they are transferred to other servers.

Some scenarios exist in which Shared mail is not used by the mail messages, regardless of the SHARED_MAIL setting. When a user saves a message at compose time, for example, the saved copy of the message is stored completely in the sender's mail file. The router has the task of separating the messages into header and body and placing the mail messages in the Shared mail database. If the message is saved at compose time, however, the router never touches it, as shown in Figure 4.18. To avoid this, suggest to your user to CC: themselves on messages they send out, rather than choosing the option of saving sent mail. This way, the router will touch the message and will have the opportunity to place the body in the Shared mail database. Only the header will be placed in the user's mail file.

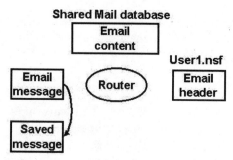

Figure 4.18   User saves message, Shared mail is not used.

Table 4.3   Shared Mail Routing Rules

SHARED_MAIL =	Recipients =	Mail Routing Path
1	1	The router places the entire message, header and content, in the user's mail file.
1	2 or more	The router places the entire message in the first user's mail file. The router then discovers the second and subsequent recipients in the TO: or CC: field. The content of the message is written to the active shared mail database. The message in the first recipient's mail file is removed and replaced with the header information only. The header information is then written to the other recipients' mail files.
2	Any	The router receives the message and immediately splits it into header and content. The content is written to the Shared mail database. The headers are written to the recipients' mail files.

NOTE
Shared mail is still used for the recipients' copies of the message in the earlier scenarios. The saved copy of the message does not use Shared mail.

Similarly, if a user edits a mail message that she has received and then saves it, that mail message will be stored in her mail file in its entirety. Again, this makes sense because the message is no longer the same as the message in the object store, and the router has not had the opportunity to touch it after changes were made.

In addition, when mail files are replicated, as in a situation where a user has a local replica of his mail file on his workstation or laptop, Shared mail is not used. For a replica of a mail file to be useful, the entire message must be available in the replica. For this reason, when a replica is created and replication of a mail file occurs, all messages will be written to the mail file. Shared mail will not be used for the messages in that mail file.

Finally, if a user uses Notes encryption to encrypt incoming messages, Shared mail will not be used. This setting is primarily used by laptop users to protect messages on their laptop from being read by a person who might obtain (and try to read) their local mail file without proper authorization. Encryption of outgoing messages does not affect Shared mail—only encryption of incoming messages.

Shared Mail Security

Using Shared mail is a security concern for some users. They worry that it would be easy for other users to have access to their mail messages. Many security features built into the Shared mail database, however, should put these fears to rest. First, the icon for the Shared mail database cannot be added to any Notes workspace. Second, the ACL of the Shared mail database is set rigidly. Only the server as a server type has access to the database. Even the workstation software on the server machine cannot access the database—only the server software. Third, Shared mail databases contain no views, and none can be added. Therefore, even if it were possible to open the database, no documents would be visible. Finally, the database is encrypted. Only the specific server ID that created the Shared mail database can access it.

Shared Mail Maintenance

Shared mail databases would become huge and unwieldy if there were no process to delete items from the database. When users delete mail messages that use Shared mail, they are only deleting the header to the message from their databases. The message body still exists in the Shared mail database, to be used by other users who have not deleted the message. Even when all users have deleted the message header from their mail files, the message remains in the Shared mail database, as shown in Figure 4.19.

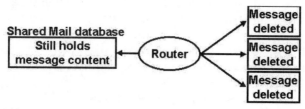

Figure 4.19   Deleting messages that use Shared mail

To remove messages that no longer have any pointers from the Shared mail database, the administrator must run the COLLECT task on the database. Collect is run by default as a 2 a.m.-scheduled server task and can also be run manually from the console by typing the following command at the server:

```
LOAD OBJECT COLLECT SHAREDMAIL.NSF
```

Any messages that are disconnected (having no pointers directed to them) from all users' mail files will be removed from the Shared mail database.

Occasionally, you may need to remove corrupted messages from the Shared mail database or remove messages after deleting or moving a user's mail file. To do this, you can run the following command at the server console:

```
LOAD OBJECT COLLECT -FORCE SHAREDMAIL.NSF
```

Be aware, however, that if a user's mail file still exists but is unavailable for some reason, you run the risk of removing the content of their mail messages. Lotus recommends that you use this command only as a last resort.

You may also need to remove headers that have become disconnected from the message content from user's mail files. To do this, you will run the following command at the server console:

```
LOAD OBJECT COLLECT USER.NSF
```

Other maintenance tasks to be run on Shared mail may include moving a user's mail file from one server to another. To do this, you must first unlink the user's mail file from the Shared mail database. Then use either File . . . Replication . . . New Replica or the operating system to make a replica of the database on the other server.

 NOTE
Why not use File . . . Database . . . New Copy? Any time you make a copy, the mail file will have a new replica ID. This function makes it a completely new database that cannot be synchronized with any other replicas of the database that may exist.

You will now relink the user's mail file to the Shared mail database on the new server, using the following command at the server console:

```
LOAD OBJECT LINK -RELINK USER.NSF SHAREDMAIL.NSF
```

You may also be asked on the exam how to limit the size or make the Shared mail database smaller. The only way to make a Shared mail database smaller or to limit its size is to create a new active Shared mail database. The easiest way to make a new active Shared mail database is to type the following command at the server console:

```
TELL ROUTER USE SHAREDMAIL2.NSF
```

SERVER CONSOLE COMMANDS RELATED TO SHARED MAIL

To enable and administer Shared mail, you will need to know some server console commands. We have already mentioned, for example, the *TELL ROUTER USE* and *LOAD OBJECT COLLECT* commands. You will also need to know commands related to linking mail files to the Shared mail database, and you should know commands that will enable you to find out whether a particular mail file is using Shared mail. Table 4.4 shows the console commands related to Shared mail. While it is probable that not every command will be tested on the exam, there will be a variety of questions that assume knowledge of the console commands.

Table 4.4   Shared Mail Server Console Commands

Command	Usage and Information
TELL ROUTER USE SHAREDMAIL.NSF	The *TELL ROUTER USE* command can be used to enable Shared mail. It will automatically change the SHARED_MAIL setting in the NOTES.INI file to two. It will also create the database link file, MAILOBJ.NSF, and the shared mail database itself, named MAILOBJ#.NSF, by default. You can also use the *TELL ROUTER USE* command to change the active Shared mail database. While the router can only write to one shared mail database at a time (the active database), users can access multiple Shared mail databases that may still be storing messages. For example, you have enabled Shared mail on your server, and your Shared mail database, MAILOBJ1.NSF, is getting too large. This can cause corruption and slower response time. At this point, you may type TELL ROUTER USE MAILOBJ2.NSF. The new database, MAILOBJ2.NSF, will be created and used as the active Shared mail database, and the pointer file, MAILOBJ.NSF, will be updated with the new information.
LOAD OBJECT INFO USER.NSF, where USER.NSF is any user's mail file	This command will tell you if the specified mail file uses an object store (Shared mail). If you have situations where there are many replicas being made, for example, this command may be useful to tell you if a particular database is using Shared mail.[1]

Command	Usage and Information
LOAD OBJECT LINK USER.NSF SHAREDMAIL.NSF	The *LOAD OBJECT LINK* command links a particular mail file (USER.NSF) or a directory (use the directory name in place of the mail filename) to the specified Shared mail database (SHAREDMAIL.NSF), as shown in Figure 4.21.
LOAD OBJECT LINK —RELINK USER.NSF SHAREDMAIL.NSF	When you use the *LOAD OBJECT LINK* command, all mail messages that have never used a Shared mail database before are linked to the Shared mail database. The —RELINK option enables all messages, including those that may have been previously linked, to be linked to the specified Shared mail database.
LOAD OBJECT UNLINK USER.NSF	Use this command when you want a particular mail file to stop using Shared mail. If you need to move a user from one mail server to another, for example, first unlink the mail file from the Shared mail database on the first server using the *UNLINK* command, as shown in Figure 4.22. You would then need to relink the user's mail file to the Shared mail database on the new server.
LOAD OBJECT COLLECT SHAREDMAIL.NSF USER.NSF	The *LOAD OBJECT COLLECT* command can be run either on the Shared mail database or on a user's mail file. Run *COLLECT* when or there are disconnected messages between the user's mail database and the Shared mail file. When all the headers pointing to a particular message in the Shared mail database have been deleted, for example, you have to run COLLECT to remove the message content from the Shared mail database.[2]

continues

Table 4.4   Continued.

Command	Usage and Information
LOAD OBJECT COLLECT—FORCE SHAREDMAIL.NSF	Adding the —FORCE option to the *LOAD OBJECT COLLECT* will remove damaged messages or any messages for which it cannot find a pointer. Run this option with caution, as messages can be deleted unintentionally if any mail files are unavailable or off-line.
LOAD OBJECT SET—ALWAYS USER.NSF	We discussed situations earlier where Shared mail is not used. If a mail file is replicated, for example, Shared mail is not used. If you want to force these mail files to use Shared mail, you can use the *LOAD OBJECT SET_ ALWAYS* command. To reverse this process, type LOAD OBJECT RESET —ALWAYS.
LOAD OBJECT SET —NEVER USER.NSF	You may occasionally want to have Shared mail enabled on a specific server but excluded for certain mail files. To exclude a particular mail file, use the *LOAD OBJECT SET - NEVER* command. To reverse this process and include a previously excluded mail file, type LOAD OBJECT RESET—NEVER USER.NSF.
LOAD OBJECT CREATE SHAREDMAIL.NSF	This command is another method of creating a new Shared mail database. If you use this method rather than the *TELL ROUTER USE X.NSF* method, you must change the path in the MAILOBJ.NSF pointer file to tell the router that this is the new active Shared mail database.

[1]The *SHOW SERVER* command is an easy way to find out whether Shared mail is enabled on a particular server, as shown in Figure 4.20.

[2]The COLLECT task is scheduled for 2 a.m. by default in the SERVERTASKSAT2 line of the NOTES.INI.

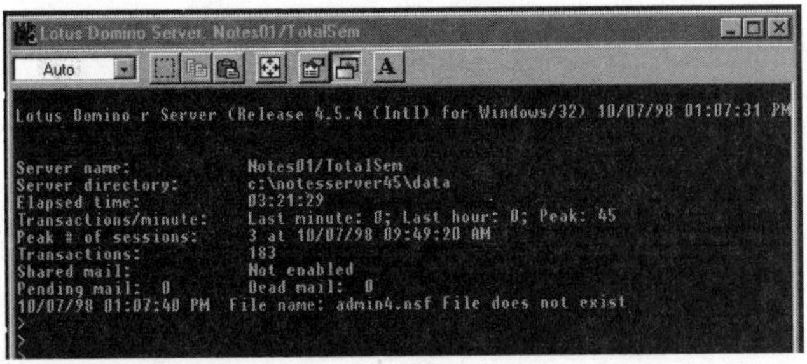

Figure 4.20 SHOW SERVER command

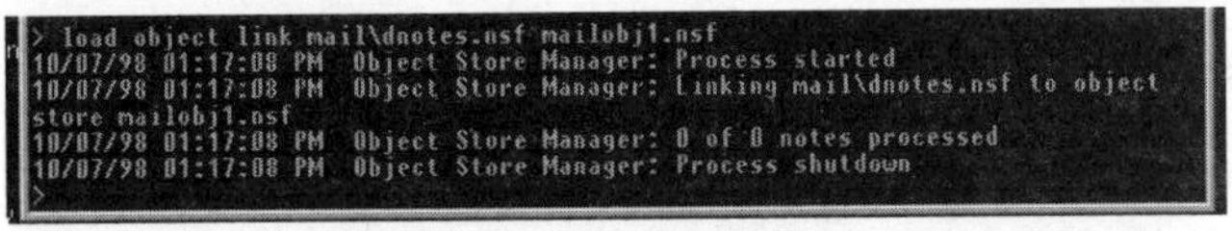

Figure 4.21 LOAD OBJECT LINK command

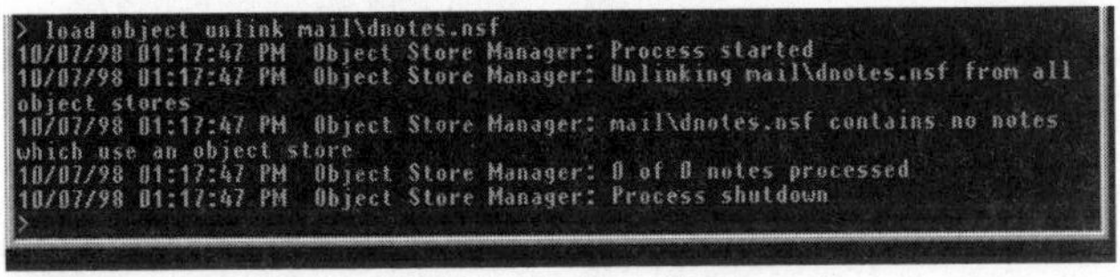

Figure 4.22 LOAD OBJECT UNLINK command

Troubleshooting Mail

The MAIL.BOX database is the first tool for mail troubleshooting, because it stores pending mail, mail awaiting transfer to users on other servers, and dead mail. When a message cannot be delivered to a recipient, it tries to send a delivery failure to the sender. This Delivery Failure report contains the original message as well as the failure message, as shown in Figure 4.23. When a delivery failure message is sent, the undeliverable message would not be considered dead, only undeliverable.

Sometimes, however, a Delivery failure report also fails to reach the original sender of the message for some reason. A message in this state, that can neither be delivered nor returned, is considered dead, as shown in Figure 4.24.

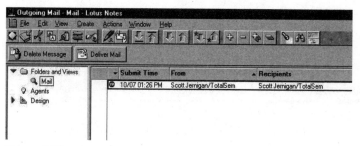

Figure 4.23   Delivery failure report

Figure 4.24   MAIL.BOX showing dead mail

You should check the MAIL.BOX regularly for dead mail or mail that has been pending for an extended period. Both situations can mean that a remote server is unavailable for some reason. This can mean that a Connection document is wrong or has been disabled, or that a server or a user's mail file is unavailable for some reason.

You can also find out whether you have dead or pending mail by typing either the SHOW SERVER or SHOW TASKS command at the server console. The *SHOW SERVER* command is better for this purpose than the *SHOW TASKS* command because of the number of items that *SHOW TASKS* will display, as shown in Figures 4.25 and 4.26.

Although MAIL.BOX is a Notes database, it is not visible in the File . . . Open Database dialog box. To open the MAIL.BOX, choose File . . . Open Database and then type in the name MAIL.BOX on the filename line. When you have the MAIL.BOX open, you will recognize dead mail by the red stop sign icon. After you have found the cause of the dead mail, you can choose the Release Dead Mail action button to deliver the message.

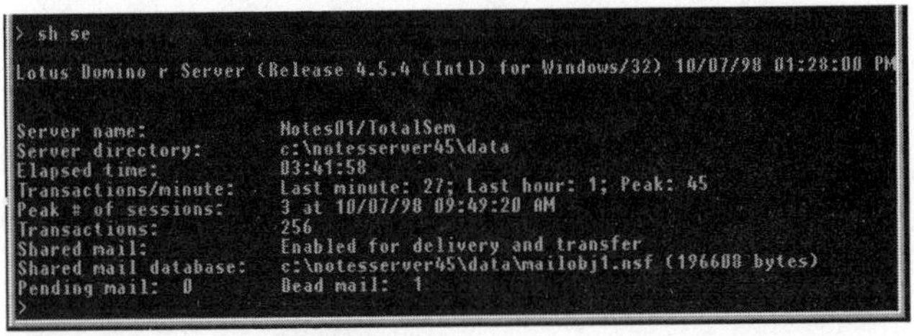

Figure 4.25 SHOW SERVER

Figure 4.26 SHOW TASKS

To help you determine the cause of dead mail, you may want to send a Mail Trace message. To send a mail trace message, open the Server Administration control panel by choosing File . . . Tools . . . Server Administration. Choose the appropriate server from the list of servers to administer. Finally, click the Mail button on the far right of the panel and choose Send Mail Trace. This button displays the Send Mail Trace dialog box, as shown in Figure 4.27. At this point, you will either type in the user's mail address or choose it from the address book. The address that you are typing is any address to which you have not been able to send mail. You can choose to receive information in the form of a delivery report from either Each Router on the delivery path or only the Last Router on path. This choice determines how many separate messages you receive, although the data is the same.

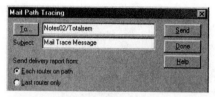

Figure 4.27 Send Mail Trace dialog box

The Mail trace message is not delivered to the recipient's mail file. Instead, the message traces the entire path from your server to the server of the recipient and back. This action is reported in a message (or messages, if you chose Each Router on Path) that you receive from the Mail Router. The message you receive shows the path taken by the message and the amount of time the message took for each hop.

NOTE

The Mail Trace feature is new for servers running Notes R4.X. Any server running an earlier version of Notes does not return a trace report.

Sometimes when troubleshooting mail routing between servers, you may not want to wait for the scheduled mail routing times. You can create a Connection document that specifies a low mail routing threshold, such as forcing mail to route with one message pending. You can also force mail to route manually from the server console by typing in the following command:

```
ROUTE SERVERNAME
```

Use this command to force the messages waiting to be transferred to the specified server to transfer immediately, instead of waiting for the scheduled routing connection time.

Review Questions

1. What does Emil need to do to enable Message-based mail on his server?

   a. Set `Message_BasedMail = 1` in the `NOTES.INI`.

   b. Type `SET CONFIG Message_BasedMail = 1` at the console.

   c. Create a configuration document in the NAB.

   d. Nothing. Message-based mail is the default.

2. What does Fran need to do to enable Shared mail on her server?

   a. Set `Shared_Mail = 1` in the `NOTES.INI`.

   b. Type `TELL ROUTER USE SHARED_MAIL` at the console.

   c. Create a configuration document in the NAB.

   d. Nothing. Shared mail is the default.

3. Which element of Notes places newly created messages in the sender's server's `MAIL.BOX`?

   a. Mailer

   b. Router

   c. Deliverer

   d. Collector

4. Which element of Notes places messages in your mail file when they are being delivered?

   a. Mailer

   b. Router

   c. Deliverer

   d. Collector

5. Which element of Notes verifies the names of the recipients in the mail messages you send?

   a. Mailer

   b. Router

   c. Deliverer

   d. Collector

6. The router checks the person documents of the sender and recipient of a message. Both the sender and recipient are on the same server. What happens to the mail message?

   a. The router looks at the server document to see when it can deliver the mail messages.

   b. The router delivers the mail directly from the sender's mail file to the recipient's mail file.

   c. The router delivers the mail from the server's MAIL.BOX to the recipient's mail file.

   d. The router looks at the MAILCONFIG= line of the NOTES.INI to see how to deliver the message.

7. The router checks the Person documents of the sender and recipient of a message. The sender and recipient are on different servers. What happens to the mail message?

   a. The router looks at the server document to see when it can deliver the mail messages.

   b. The router delivers the mail directly from the sender's mail file to the recipient's mail file.

   c. The router delivers the mail from the server's MAIL.BOX to the recipient's mail file.

   d. The router delivers the mail from the server's MAIL.BOX to the recipient's server's MAIL.BOX.

8. The router checks the Person documents of the sender and recipient of a message. The sender and recipient are on different servers. SHARED_MAIL is set equal to 1. Is Shared mail used to deliver the message?

   a. Yes

   b. No

9. If the router finds the same NNN in the server documents of the sender and the recipient, how is mail delivered?

   a. According to the schedule in the server document

   b. According to the schedule in the Connection document

   c. According to the schedule in the Configuration document

   d. Immediately and automatically

10. If the router finds a different NNN in the server documents of the sender and the recipient, how is mail delivered?

   a. According to the schedule in the server document
   b. According to the schedule in the Connection document
   c. According to the schedule in the configuration document
   d. Immediately and automatically

11. How many Connection documents are required for complete mail routing between servers?

   a. Zero
   b. One
   c. Two
   d. Three

12. Which element of Notes places mail messages in the Shared mail database?

   a. Mailer
   b. Router
   c. Deliverer
   d. Collector

13. What is the default routing cost when the connection to a remote server is made using a LAN?

   a. One
   b. Two
   c. Three
   d. Five

14. What is the default value for the Routing threshold?

   a. One
   b. Two
   c. Three
   d. Five

15. What is the main purpose of using Shared mail?

16. If Shared mail is enabled for delivery only and you send a message to one person, where is the body of the message stored?

    a. SHAREDMAIL.NSF
    b. MAILOBJ.NSF
    c. The user's mail file
    d. NAMES.NSF

17. If Shared mail is enabled for delivery and transfer, and you send a message to one person, where is the body of the message stored?

    a. SHAREDMAIL.NSF
    b. MAILOBJ.NSF
    c. The user's mail file
    d. NAMES.NSF

18. If Shared mail is enabled for delivery, and you send a message to two people, when is the message body written to SHAREDMAIL.NSF?

    a. As soon as the router sees the first recipient
    b. As soon as the router sees the second recipient
    c. After the router writes the message to both users' mail files
    d. After the router writes the header to both users' mail files

19. Which server command is better for verifying the Shared mail setting?

    a. SHOW SERVER
    b. SHOW TASKS

20. Where can you look to find out whether there are dead mail messages?

    a. Server console using SHOW TASKS
    b. Server console using SHOW SERVER
    c. Server console using SHOW DEAD
    d. MAIL.BOX

21. If you plan to move Scott's mail file from server NOTES01 to server NOTES02, what must you do first, assuming you are using Shared mail?

    a. LOAD OBJECT LINK -TRANSFER SCOTT.NSF NOTES02.NSF
    b. LOAD OBJECT LINK - RELINK SCOTT.NSF MAILOBJ2.NSF
    c. LOAD OBJECT UNLINK SCOTT.NSF
    d. LOAD OBJECT COLLECT -FORCE SCOTT.NSF

22. Where do you go to send a Mail Trace message?

    a. File . . . Tools . . . User Preferences . . . Mail . . . Mail Trace
    b. File . . . Tools . . . Server Administration . . . Mail . . . Mail Trace
    c. File . . . Tools . . . Action . . . . . .Send Mail Trace
    d. MAIL.BOX database

23. If you are saving your sent messages automatically, and Shared mail is enabled for delivery and transfer, how many copies of the message are stored on your server?

    a. One
    b. Two
    c. Three
    d. Four

Answers

1. d. Nothing. Message-based mail is the default.

2. Either a or c. You can enable Shared mail by setting the SHARED_MAIL = line in the NOTES.INI to one or two by creating a configuration document, or by using the TELL ROUTER USE SHAREDMAIL.NSF command at the server console.

3. a. Mailer

4. b. Router

5. a. Mailer

6. c. The router delivers the mail from the server's MAIL.BOX to the recipient's mail file.

7. d. The router delivers the mail from the server's MAIL.BOX to the recipient's server's MAIL.BOX.

8. b. No. When SHARED_MAIL = 1, it is used for delivery only. To use SHARED_MAIL for delivery and transfer, set SHARED_MAIL = 2.

9. d. Immediately and automatically

10. b. According to the schedule in the Connection document

11. c. Two Connection documents are required, because mail routing is a one-way process. (You can use zero Connection documents if your servers are in the same NNN.)

12. b. Router

13. a. One

14. d. Five

15. Saving space on the mail server

16. c. The user's mail file; Shared mail is not used.

17. a. SHAREDMAIL.NSF

18. b. As soon as the router sees the second recipient

19. a. SHOW SERVER. Either command contains the information, but it is easier to read with the SHOW SERVER.

20. a, b, and c are correct, although b and c are the best answers.

21. c. LOAD OBJECT UNLINK SCOTT.NSF

22. b. File . . . Tools . . . Server Administration . . . Mail . . . Mail Trace

23. b. Two

CHAPTER 5

Notes Replication

Objectives

After reading this chapter, you should be able to answer questions based on the following objectives:

- Define replication.
- Create a replica of a database.
- Define and use the different replication types.
- Create a server-to-server replication schedule.
- Force replication using the server console.
- Enable multiple replicators.
- Use the workstation replicator page to create a workstation-to-server replication schedule.
- Use a location document to create a workstation-to-server replication schedule.
- Force replication from a workstation.
- Troubleshoot replication problems.

- Define the purge interval for documents deleted from a replica database.
- Define and troubleshoot replication conflicts.

Introduction to Replication

Replication is the process of exchanging data and synchronizing information between Notes databases. Replication enables us to have multiple duplicates of a database on multiple servers and workstations, all altered by different people, and still maintain uniformity in the content of that database. This grants flexibility to the work force, because they can easily be in different locations. Replication gives these benefits without any sacrifices.

Replication has several requirements. First, you must have a replica of a database on more than one Notes machine (server or workstation). Second, you must decide on the type of replication that you want to use, such as pull, push, pull-pull, and pull-push. Third, you must determine whether the replication will happen between server and server or server and workstation. Each requires specific settings and tweaking, but each also shares many common characteristics. Server-to-server replication requires resolution of the following issues:

- A Replicator task must be running on the server.
- Replication must be initiated.
- A communication pathway between the servers must exist, including physical connectivity, shared network protocol, proper topology (or if there is no shared protocol, a Passthru server must be used).
- Other factors must be resolved, including authentication, ACLs, database replica IDs, and replication conflicts.

Workstation-to-server replication has almost the same requirements as server-to-server replication, but workstation-to-server replication requires different steps for initiation and scheduling. Finally, you must have some idea of troubleshooting in order to service replication processes. This chapter covers each of these issues in detail.

Creating Replicas

Replication requires a replica, not a copy, of a database on more than one machine. A *replica* of a database is an exact duplicate of the original database, including an identical replica ID. A *copy* of a database, on the other hand, duplicates the original database, but has a completely different replica ID. The procedures to create a replica or a copy vary greatly and are discussed in the following paragraph. To create a copy, use File . . . Database . . . New Copy. To verify that the replica ID is different when you make a copy, right-click on the database icon and choose Database Properties (or choose Database Properties from the File menu). In the Database Properties InfoBox, click on the Information tab to verify the replica ID, as shown in Figure 5.1.

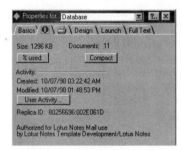

Figure 5.1 Replica ID

You can create a replica of a database either through the operating system copy feature or internally in Notes. If you use Windows (NT, 95, or 98), you can use Windows Explorer or My Computer to select the .NSF file and make a copy onto your workstation or server. Be aware that if you use this method the database cannot be in use. If you are copying from a server, this may require the Notes server program to be shut down. An operating system copy, unlike an internal Notes copy, duplicates a database exactly—including the replica ID.

Inside Notes, you can create a replica by selecting File . . . Replication . . . New Replica from the menus. If you have an icon for the database already added to your workspace, select the icon before choosing New Replica from the menu. Notes assumes that you want to make a replica of the selected database. The New Replica dialog box is displayed with most of the fields already completed, as shown in Figure 5.2. The server field should be Local, because you are making a replica of the database on your local machine (either a server or a workstation). If you are making a replica on a server from a workstation database, choose the correct server in this field.

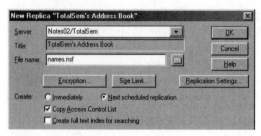

Figure 5.2 New Replica dialog box

 NOTE
Your name must be in the Create Replica Databases field of the server document to create new replicas on a server.

The database title and filename fields are completed from the information provided by the selected database. If you choose to create the replica immediately, the workstation's replicator immediately creates the replica and initializes the design and documents. This means that Notes copies the design and documents and makes the views ready for immediate use.

If you choose to create the replica at the next scheduled replication, only a replica stub is created on the workspace, as shown in Figure 5.3. You will not be able to open a replica stub before it has been initialized through a complete replication, because it does not yet have any views. If you try to open a replica stub, you receive the error shown in Figure 5.4.

Figure 5.3 Replica stub on the workspace

Figure 5.4   Replica not yet initialized error

If you do not already have an icon for the database on your workspace, you have two choices. You can add the icon and then proceed as described earlier, or you can create a replica without an icon.

To add a database icon to your workspace, choose File . . . Open Database . . . , or right-click the workspace and choose Open Database . . . . The Open Database dialog box is displayed, as shown in Figure 5.5. Select the server where the database resides and then select the database from the list. Subdirectories under the \NOTES\DATA directory are shown at the bottom of the list. If you just want to add the icon without opening the database, click Add Icon. To both add the database icons and open the database, click Open.

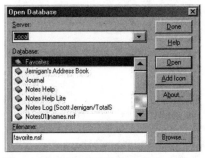

Figure 5.5   Open Database dialog box

To create a replica when you do not have an icon for the database on your workspace, make sure you do not have any icons selected and choose File . . . Replication . . . New Replica. The Choose Database dialog box, which looks almost exactly like the Open Database dialog box, is displayed. This enables you to select the database that you want to use as the basis for the replica. Select the appropriate server and database, and then click the Select button, as shown in Figure 5.6.

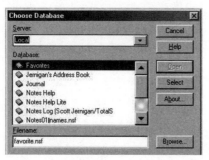

Figure 5.6 Select a database

After you click the Select button, the New Replica dialog box is displayed with most of the fields already completed from the database you selected. The rest of the process is the same as described earlier. The server field should be Local, because you are making a replica of the database on your local machine, either a server or a workstation. If you are making a replica on a server from a workstation (or another server), choose the correct server in this field.

After you have created replicas on your servers or workstations, you will want to be able to initiate and schedule replication. These processes are discussed in the rest of the chapter.

Types of Replication

There are four types of replication: pull-only, push-only, pull-push, and pull-pull. Be aware that although we will discuss compound replications, where the synchronization between the servers goes both ways, replication itself is a one-way process. The idea of *two-way* replication only indicates that a secondary replication is initiated immediately following another replication. The System Administration I exam tests your familiarity with these replication types.

Push Replication

Push-only replication causes a one-way synchronization of data. The initiating server calls or connects to the receiving server. The initiating server then pushes the changes in its databases to the receiving server. Another way to say this is that the source server— the server that initiated the call—writes its changes to the target (or receiving) server.

If you have a server in your main office and a server in a satellite office, for example, you may allow only administrators at the main office to make changes to the Public NAB. When the administrators make these changes, however, the other replicas of the NAB need to be updated. In this case, a push-only replication of the NAB from the main office to the satellite office would be appropriate, as you can see in Figure 5.7.

Initiating   Receiving

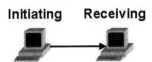

Figure 5.7 Push-only replication

Pull Replication

Pull-only replication is also a one-way synchronization. In this scenario, the initiating server calls the receiving server, exactly as it would in push-only replication. In this case, however, the initiating (or calling) server pulls the changes from the receiving (or remote) server's databases. In pull replication, the initiating server becomes the target server, and the remote server becomes the source server.

In our previous example of a main office server and a satellite office server, the satellite office server may need to control replication because of a communications requirement, such as a dial-up modem. Continue to assume that the changes to the Public NAB are made by administrators at the main office. The satellite office server, however, must call or connect to the main office server. It then initiates a pull-only replication with the main office server and pulls all the NAB changes down, as you can see in Figure 5.8.

Initiating   Receiving

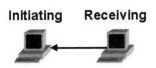

Figure 5.8 Pull-only replication

Pull-Push Replication

Pull-push replication is the most common type of replication because it is the default, both when you create a replication Connection document and when you initiate replication at the server console. While the name may suggest otherwise, pull-push replication is one-way replication, as we discussed earlier. It is, however, a compound replication. This means that two one-way replications occur in sequence. The initiating server makes the call to the remote server. The initiating server begins a pull replication with the remote server and pulls to itself the changes that exist in the remote server's databases. Then the initiating server immediately begins pushing its changes to the remote server. The initiating server controls both of the replications—it does all the work and makes all the changes. The remote server's replicator task remains idle the entire time. The initiating server begins as the target of the changes from the remote (or source) server. When it begins to push changes, it becomes the source server—and the remote server becomes the target.

In our scenario, we have the main office server and the satellite office server. In this case, we assume that there are administrators making changes to the Public NAB at both locations. The remote administrators may be given the ability to register users, for example. When changes are being made in both replicas, a full synchronization of the database is necessary. If you initiated a pull-push replication from the main office server, it would call the satellite office and pull the changes from the NAB at the satellite office. The main office server would then initiate a push of the changes contained in its replica to the satellite office. At the end of the replication, the two databases would be completely synchronized. This process is shown in Figure 5.9.

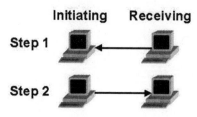

Figure 5.9 Pull-push replication

Pull-Pull Replication

Pull-pull replication is also a compound replication that uses two one-way replications to synchronize the databases fully. In this case, the initiating server makes the connection with the remote server and begins pulling the remote server's changes. When the first pull replication is completed, the replicator task on the initiating server tells the remote server to begin a pull replication. This is sometimes called tagging. The remote server then initiates a pull replication to complete the cycle. Each server's replicator task works in this manner.

In our scenario of the main office server and the satellite office server, you might choose to use pull-pull replication if either both replicators or both servers are fairly busy. Pull-pull replication shares the work more equally between both servers. The main office server controls the replication, initiating a pull-pull replication with the satellite office and pulling the satellite office server's changes. When the changes have all been replicated to the main office server, the main office server's replicator tags the satellite office server's replicator task and then becomes idle. The satellite office server's replicator task then pulls the changes from the main office server, completing the replication. This process is shown in Figure 5.10.

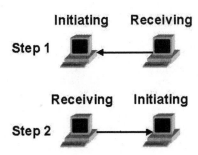

Figure 5.10 Pull-pull replication

Server-to-Server Replication

Replication of a database from one server to another in a Notes environment requires resolution of several factors, including a Replicator task, initiation, connectivity, and other factors. Servers run a replicator task automatically if they have the ServerTasks=Replica ... line in their NOTES.INI file. A server must run a replicator task to replicate a database. Notes places the Replica task in the ServerTasks line by default, to ensure that the replicator task runs. For most servers, you will allow replication to run on a regularly scheduled basis. This requires you to leave the Replica task in the NOTES.INI. You might remove the Replica task if your server will not be replicating regularly and you need to free up server resources. Replicator tasks remain idle unless needed but require 3MB of RAM even in that state.

Initiating Replication

Server-to-server replication is initiated in two different ways: either at the server console or with a replication schedule. Server console initiation is less common, because it requires an administrator to type the commands at the console. More often you will use a replication schedule to allow the server Connection documents to initiate replication automatically.

Initiating Replication at the Server Console

If changes made to a database need to be replicated immediately, an administrator may not want to wait for the scheduled replication to occur. In this case, you may choose to initiate replication manually at the server console. Three different commands initiate replication at the server console:

- PULL servername databasename
- PUSH servername databasename
- REPLICATE servername databasename

Servername is the name of the server with which you wish to replicate. *Databasename* is either the name of the specific database you wish to replicate (relative to the data directory path), or a directory you wish to replicate (again, relative to the data directory). You may also choose to leave out the *databasename* parameter, if you wish to replicate all the databases in the \NOTES\DATA directory.

The PULL and PUSH commands initiate pull-only or push-only replication, while the REPLICATE command initiates pull-push replication. You cannot initiate pull-pull replication from the server console.

NOTE
Another way to describe replication is to use the words *source database* and *target database*. The target database is always the database that is receiving changes. The source database is always

the database that contains the changes. This can be confusing when combined with remote servers and the concept of initiating. Pay close attention to the flow of information between servers in examples or questions. The source and target databases are always determined by the flow of information.

After you initiate replication, a series of factors affect whether the replication occurs. First, the initiating server must find a path to the indicated remote server by locating a direct connection or by using a Connection document in the NAB. After finding the remote server, the initiating server must verify that it can communicate with and use that remote server. The replicator must then verify that a replica database exists on the remote server. The initiating server must have appropriate access to the remote database and all the documents in the remote database. These and other factors affecting replication are described more fully later in this chapter.

Server Connection Documents and Replication Settings

Most of the time you will want replication to initiate automatically, so you will create server Connection documents with replication schedule times enabled. You can use any of the four types of replication in server Connection documents. In addition, you can use multiple Connection documents to create a more complete replication schedule.

You may want to use pull-only replication from a central server to bring in changes from many other satellite offices, for example. You might then want that central server to push out the collected changes to all the satellite offices later in the day. To create this scenario, you would use two replication Connection documents per server: one that uses pull-only and one that uses push-only.

In Figure 5.11, you can see a server Connection document with the fields pertinent to replication completed. The following paragraphs describe the server Connection document and the replication schedule.

To create a server Connection document, go to the Servers . . . Connections view. Choose the Add Connection action button. The Connection document is similar to the Connection document for a Mail Routing connection, discussed in Chapter 4, "Notes Mail." In this case, however, you will place Replication in the Tasks field, to make the Connection document a replication document. The fields in the Replication Connection document are described in Table 5.1.

Table 5.1 Replication Connection Document

Field Name	Value	Notes
Connection Type	This field name describes the manner in which the two servers are connected: for example, LAN, dial-up modem, and Passthru server. Choose the appropriate type for your servers. Please note that the value in this field changes the other fields.	You can have multiple Connection documents between any two servers. You may want to use different connection types in case of network outages.
Source Server	Type the hierarchical name of the server that originates the connection. The server that you enter here will be the initiating server in all the examples.	By default, this is the server on which you create the document. For example, if you have the NAB open on ServerA when you create the Connection document, ServerA will be the source server automatically. You can (and often must) change this field.
Source Domain	The source domain is the Notes domain in which the source server exists.	Empty by default
Use the ports	Define the port for this connection to use. In the case of a LAN Connection document, this relates to the protocol: for example, TCP/IP or IPX/SPX. For a dial-up modem connection type, this is usually COM1.	

Field Name	Value	Notes
Usage Priority	You can set a low-usage priority if you want a particular connection to be used only as a last resort. For example, if two servers were connected over a bridged/routed network, this would probably be your first choice of methods to connect for replication. You might also have a Connection document for modems on both servers. The dial-up modem connection document would use Low Priority.	Normal or low; default is Normal.
Destination Server	The destination server is the hierarchical name of the server to which you are connecting. In the examples used earlier, this is the remote server.	
Destination Domain	The destination domain is the Notes domain in which the destination server exists.	
Schedule	Choose whether the schedule is enabled or disabled. At times you may need to disable a certain connection, for example when a server is off-line, but you may not want to delete the Connection document.	

continues

Table 5.1 Continued

Field Name	Value	Notes
Call at times	This phrase indicates the times at which the server should make contact with the destination server. This can be specific times or ranges of time.	Default is 8 A.M. through 10 P.M.
Repeat Interval	The value in this field describes how often the source server will attempt to contact the destination server. If only one time or one call is desired, this field can be left blank.	Default is 360 minutes.
Days of Week	Choose the days of the week that this Connection document is valid. You may choose to have different connection times for the weekends, for example.	The week begins on Sunday and goes through Saturday.
Tasks	Your choices in this field include Mail Routing, Replication, X.400 Mail Routing, SMTP Mail Routing, and cc:Mail Routing.	Choose Replication in this case. Default is both Replication and Mail Routing.
Replicate Databases of ___ Priority	This field determines the databases that are replicated, based on replication priority level. If you only want databases of High priority to be replicated on a particular schedule, you can indicate this using this field.	Low-, Medium-, and High-priority databases are the default. Your other choices are High-priority and Medium- and High-priority. To set a replication priority for a database, choose the Other panel of the Replication settings dialog box for the database. This option is discussed later in this chapter.

Field Name	Value	Notes
Replication Type	This field determines the type of replication to be used for this Connection document.	Choices include pull-only, push-only, pull-pull, and pull-push. The default replication type is pull-push.
Files/Directories to Replicate	In this field, choose the files (databases or directories) that should be replicated on this schedule. If you leave the field blank, all databases in these three categories will be replicated: ■ Databases with replicas on the receiving server in the \NOTES\DATA directory ■ Databases with replicas on the receiving server in a subdirectory of the data directory ■ Databases with replicas on the receiving server that use a directory link located in the data directory	The default is blank, forcing all databases with replicas to replicate. Usually it is a good idea to replicate files like NAMES.NSF, ADMIN4.NSF, and STATREP.NSF, on a separate Connection document to ensure that they are being replicated.
Replication Time Limit	This field indicates the number of minutes the replication can take before the server automatically stops the replication.	By default, this field is blank. Lotus recommends using this field carefully. If there is a slow connection, replication may be stopped before any real progress is made.[1]

---

[1] If there is a replication time limit set, and a server crashes during replication so that the replication does not complete, the replication will begin where it left off when it starts again.

SERVER CONNECTION: Notes01/TotalSem

Basics

Connection type:	Local Area Network	Usage priority:	Normal
Source server:	Notes01/TotalSem	Destination server:	
Source domain:		Destination domain:	
Use the port(s):		Optional network address:	

Choose ports

Scheduled Connection		**Routing and Replication**	
Schedule:	ENABLED	Tasks:	Replication
Call at times:	08:00 AM - 10:00 PM each day		
Repeat interval of:	360 minutes		
Days of week:	Sun, Mon, Tue, Wed, Thu, Fri, Sat	Replicate databases of:	Low & Medium & High priority
		Replication Type:	Pull Push
		Files/Directories to Replicate:	totalnames.nsf (all if none specified)
		Replication Time Limit:	minutes

Figure 5.11 Server connection document with a replication schedule enabled

The Connection documents that you create for replication should be defined by the replication topology you have chosen to implement. The next section describes communications pathways, including the common replication topologies.

Communication Pathways

Once you have the replicator task running and replication is initiated with the correct type of replication, you must select the proper topology. You must have the proper physical connectivity, including network settings and protocols. If your servers do not share the same protocol, you must use a Passthru server to define the proper topology for the job. Topologies and Passthru servers are discussed in the next paragraphs.

Replication Topologies

Four basic replication topologies exist, as discussed in Chapter 4, "Notes Mail": hub and spoke, end-to-end, ring, and mesh. Topology describes where servers are located and how they are connected. The basic reason for initiating a replication topology is to control and predict how information is distributed within your organization.

In *hub and spoke* topology, one server (the hub) schedules, initiates, and controls all replication. This is the best way to control data and ensure that all data is accurate. You may have a corporate office in Houston, for example, and smaller offices all over the world. You might choose to have the hub server in Houston be the central hub for your enterprise. In this case, the hub calls the other servers (or initiates the connection over a LAN or WAN) and initiates replication. All the other servers are the spokes. They only pass information to other databases to and from the hub when they are called by the hub. The hub and spoke can be expanded to have multiple levels, in which case it is called a *binary tree* topology. Hub and spoke topology will use pull, push, and pull-push replication in most cases but will use only pull-pull replication if there is a need to reduce work done by the hub replicators. Hub and spoke is the most common replication topology in enterprise environments because it is the most efficient and organized. An example of hub and spoke topology is shown in Figure 5.12.

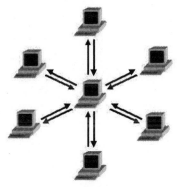

Figure 5.12   Hub and spoke topology

End-to-end replication topology means each server is connected in a line, as shown in Figure 5.13. Each server talks to the server ahead of it and after it. In this chain, for example, the server in our Houston office calls the server in our Dallas office. The Dallas server calls the Kansas City office. The chain would then reverse on itself, and Kansas would call Dallas, and so on.

A *Ring* topology is similar to the end-to-end topology. When the last server is reached in the line, however, it simply circles back to the beginning, as shown in Figure 5.14. So in our example, after Dallas calls Kansas City, Kansas City would turn around and call Houston.

Figure 5.13 End-to-end topology

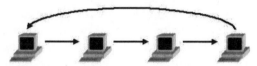

Figure 5.14 Ring topology

In *Mesh* topology, all servers connect directly to all other servers, as shown in Figure 5.15. This can be a difficult topology to administer because of its complex and disorganized nature—and is not recommended.

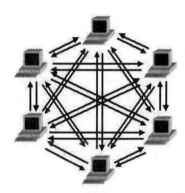

Figure 5.15 Mesh topology

Two basic resource management issues—bandwidth and communication resources and scheduling replications—influence the type of replication topology you will select for your organized Notes environment. Bandwidth and communication resources are not within the purview of Notes System Administration but certainly must be considered. How fast are your links? Are they all 56K, or are some 56K and others 256K? The System Administration I exam does not test these concepts, but you must think about these items when administering a Notes environment.

To schedule replication properly, you must ensure that information gets where it needs to go in a timely and error-free fashion. Users must not miss important information, and servers always need to be coordinated. Time zones and multiple schedules complicate these issues. If your servers are in different time zones, take into consideration the servers' peak and low times—as well as the peak and low times of your own servers. Also, plan your replication schedule so that your replicator is not overworked and you do not lose changes in the replication. When creating your replication schedule, the NAB is the most important database to consider. After determining the best replication topology and schedule for the NAB, you should consider any other databases that are used for your organization's business.

In your replication topology, your server Connection documents schedule one of the types of replication with another server or set of servers. In the most popular type of topology, hub and spoke, the hub contacts all the servers and pulls the changes to the hub from the spokes. After all of the scheduled pulls, the hub again contacts the spoke servers and then pushes the amassed changes out to the spokes.

For example, you may have a corporate office in Houston and smaller offices all over the world. You might choose to have the hub server in Houston be the central hub for your enterprise. In this case, the hub will call the other servers (or initiate the connection over a LAN or WAN), and initiate a pull-only replication. The changes made in the databases that are being replicated—for example, NAMES.NSF, STATREP.NSF, and ADMIN4.NSF—are pulled up to the hub. The hub acts as the repository for all the changes. Later, the hub calls each of the spoke servers and pushes the combined changes out to the remote databases. This replication topology sample is shown in Figure 5.16.

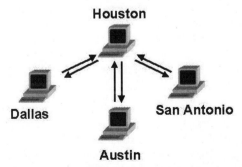

Figure 5.16 Sample replication technology

Multiple Replicators

The server replicator task only services one replication request at a time. While the replicator is servicing that request, it can also keep track of up to five other replication requests. This memory is called the *replicator queue*. When the replicator queue is full, other replication requests will be dropped and the requested replication will not occur. One way to avoid dropped replication requests on a busy server is to enable more than one replicator task on the server. Obviously, multiple replicators will require more server resources, in terms of memory, server input/output, processor, and network bandwidth. Each replicator uses a minimum of 3MB of RAM, even when idle.

To enable multiple replicators, change the REPLICATORS line in the server's NOTES.INI file. You can do this either directly in the NOTES.INI file by adding a line with REPLICATORS=*n* (where *n* is the number of replicators you would like to enable), or by creating a server configuration document in the NAB. The latter is the more recommended method.

To create a server configuration document, open the NAB to the Servers . . . Configurations view. Click the Add Configuration action button. This opens a Server Configuration form, as shown in Figure 5.17.

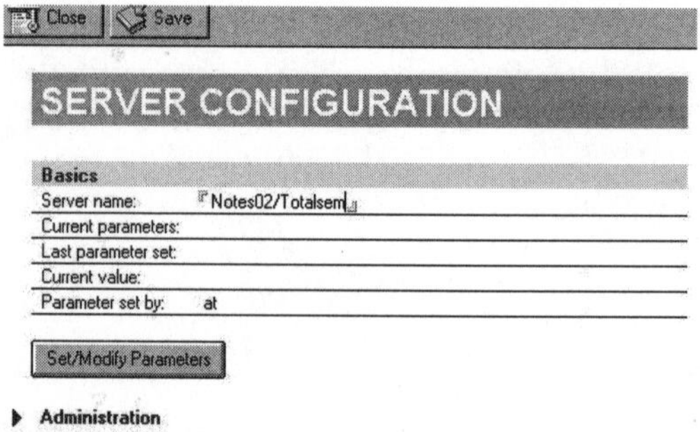

Figure 5.17 Using the Server Configuration form to enable multiple replicators

Fill in the Server name with the appropriate server in hierarchical format. If you want the configuration to apply to all servers in the domain, place an asterisk in the field. Click the Set/Modify Parameters button. In the Server Configuration Parameters dialog box, fill in the Item field with the REPLICATORS parameter, as shown in Figure 5.18. In the value field, type the number of replicators you want to enable.

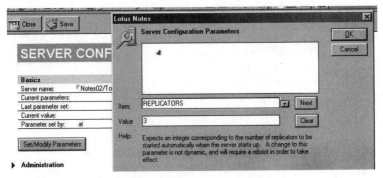

Figure 5.18 Server Configuration Parameters dialog box

You can verify that you have enabled multiple replicators by looking at the server console after a server restart. As the tasks start, you should see multiple instances of the replicator task. If you type in the command Show Tasks at the server console, you should see multiple replicator tasks in the server task list.

NOTE
You have the ability to enable a maximum of 10 replicators. In most environments, however, you will enable no more than two or three replicators.

Replication Passthru

Sometimes servers in your environment are not directly connected. A common situation is when some servers use the TCP/IP protocol, for example, and other servers use the IPX/SPX or NetBEUI protocol. When you have servers that cannot communicate directly, you must create some type of communication path—or no data can be passed between them. Server Passthru can be used for replicating with a server to which you do not have a direct connection or access. *Passthru* is a new Notes R4 feature that enables Notes workstations and servers to use a designated server as a bridge to connect to another server that does not share a common protocol with the workstation or server. This section describes how to configure and use Server Passthru for replication.

To understand the need for Server Passthru, consider the following example. There may be three servers in your environment. One server (MainServer) uses both TCP/IP and NetBEUI. The other servers are each running one of the two protocols, as shown in Figure 5.19. This means that the other servers, one (ServerA) running TCP/IP and one (ServerB) running NetBEUI, cannot communicate directly. If they are in the same domain, however, the NAB and other databases will have to be updated. In this scenario, you could simply use hub and spoke topology and use the server with both protocols as the hub. You may have certain databases, however, that you do not wish to place on the hub. In that case, you would use the hub (Main-Server) as a Passthru server for the other two servers.

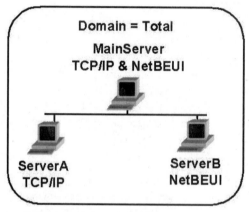

Figure 5.19 Server Passthru diagram

Server Passthru requires you to create and edit two types of documents in the Public NAB. First, you need to create a regular Connection document to schedule replication with a particular server. Second, create another Connection document specifying which server to use for the Passthru connection. Finally, you need to edit your server documents to enable Passthru.

The Connection document that you create to schedule replication is the same as the replication Connection document we created earlier. Fill in the source and destination servers and the other replication fields, as shown in Figure 5.20. This document specifies which databases to replicate with which server and on what schedule.

You then need to create a Connection document to specify the Passthru server, so that your disconnected servers know how to communicate with each other. ServerA is trying to communicate with ServerB, but they do not share a protocol. Therefore, to communicate, they will need to use the MainServer as the Passthru. This is represented in the NAB by the Connection document, shown in Figure 5.21.

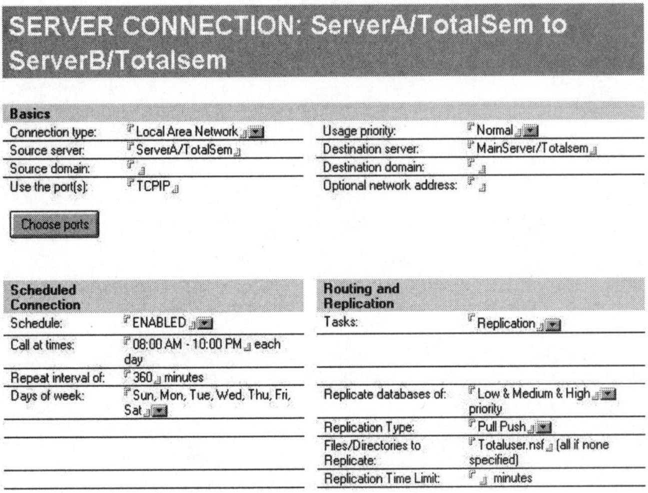

SERVER CONNECTION: ServerA/TotalSem to ServerB/Totalsem

Basics

Connection type:	Local Area Network	Usage priority:	Normal
Source server:	ServerA/TotalSem	Destination server:	MainServer/Totalsem
Source domain:		Destination domain:	
Use the port(s):	TCPIP	Optional network address:	

Choose ports

Scheduled Connection

Schedule:	ENABLED
Call at times:	08:00 AM - 10:00 PM each day
Repeat interval of:	360 minutes
Days of week:	Sun, Mon, Tue, Wed, Thu, Fri, Sat

Routing and Replication

Tasks:	Replication
Replicate databases of:	Low & Medium & High priority
Replication Type:	Pull Push
Files/Directories to Replicate:	Totaluser.nsf (all if none specified)
Replication Time Limit:	minutes

Figure 5.20   Replication Connection document

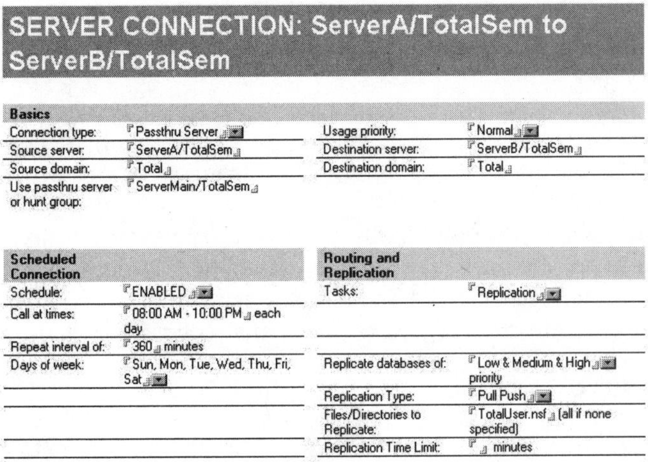

SERVER CONNECTION: ServerA/TotalSem to ServerB/TotalSem

Basics

Connection type:	Passthru Server	Usage priority:	Normal
Source server:	ServerA/TotalSem	Destination server:	ServerB/TotalSem
Source domain:	Total	Destination domain:	Total
Use passthru server or hunt group:	ServerMain/TotalSem		

Scheduled Connection

Schedule:	ENABLED
Call at times:	08:00 AM - 10:00 PM each day
Repeat interval of:	360 minutes
Days of week:	Sun, Mon, Tue, Wed, Thu, Fri, Sat

Routing and Replication

Tasks:	Replication
Replicate databases of:	Low & Medium & High priority
Replication Type:	Pull Push
Files/Directories to Replicate:	TotalUser.nsf (all if none specified)
Replication Time Limit:	minutes

Figure 5.21   Server Passthru Connection document

Most of the fields are completed the same as in any other Connection document, with the exceptions in Table 5.2.

Table 5.2 Completing the Server Passthru Connection Document

Field	Value and Information
Connection Type	Passthru Server
Use Passthru Server	MainServer

Finally, you need to edit the server documents for all the servers involved in the Passthru connection. By default, Passthru use is not enabled. Complete the server document fields as described in Table 5.3.

▼ Restrictions

Server Access	Who can -		Passthru Use	Who can -
Only allow server access to users listed in this Address Book:	⌐ No		Access this server:	⌐ Passthru_Users, LocalDomainServers
Access server:	⌐ */TotalSem		Route through:	⌐ Passthru_Users, LocalDomainServers
Not access server:	⌐		Cause calling:	⌐ Passthru_Users, LocalDomainServers
Create new databases:	⌐ Admin		Destinations allowed:	⌐
Create replica databases:	⌐ Admin			

Figure 5.22 Sample server document allowing Passthru

NOTE
Only servers that use Notes R4.X and higher can initiate Passthru or be designated as the Passthru server. Servers that use Notes V3.X can only be the destination of a Passthru connection.

A sample server document with the necessary fields completed for Passthru is shown in Figure 5.22. After the Connection and server documents are created and edited, Passthru will occur automatically and transparently, according to the replication schedule.

NOTE
We discuss the topic of Passthru again in Chapter 6, "Administration Tools and Tasks," as it relates to the administrative tasks required for remote users to use Passthru.

Table 5.3   Passthru Options in the Server Document

Field	Value and Information
Access this server	This field defines who can access this server using Passthru. You must include both servers and users. If it is blank, no one is allowed to access this server using a Passthru server.
Route through	This field defines who can use this server to route to another server. In our example, MainServer would have to allow both ServerA and ServerB to route through. If it is blank, no one is allowed.
Cause Calling	This field is officially defined as who can cause this server to place a call to another server. Any server seeking to use this server to replicate with another server, however, must be in this list. In our example, MainServer must allow ServerA and ServerB in this list. If it is blank, no one is allowed.
Destinations Allowed	This field defines remote servers that can be routed to using this server for Passthru. If it is blank, all remote servers are allowed. If you want to restrict which servers can be contacted using any particular Passthru server, you would complete this field.

Factors Affecting Replication

Many factors at all layers of the Notes environment affect replication. These factors include the following:

- Server factors, such as authentication, server access lists, server replication schedules, and number of replicators
- Database factors, such as the ACL, replica ID, replica ID cache, Replication settings, replication history, and deleted documents

■ Document factors, such as form access lists and merge replication conflicts

These factors can cause replication not to occur or to occur in an undesired manner. A good administrator will look at each of these factors when setting up and troubleshooting replication. We discuss each of the factors affecting replication here. To visualize the factors that affect replication, refer to Figure 5.23.

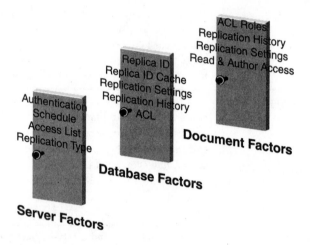

Figure 5.23   Factors affecting replication

EXAM TIP
The exam will probably not ask questions about all of these factors, but you will be expected to know how to determine what will be replicated and why in any given scenario.

Server Factors

The first layers of factors affecting replication are the server factors. We already discussed the concepts of the replication schedule and the replicator task, both of which are server factors. We can also add into the mix the replication type, as discussed earlier. Finally, the capacity to authenticate with and to access the server are important server factors. As we discussed in Chapter 3, "Notes Security," the server attempting to request a replication must be able to authenticate with the destination server. The two must share some certificate that enables them to authenticate before replication can occur.

In addition, the server access list in the server document must enable the initiating server to access the receiving server. If the initiating server is not in the remote server's access list or is in the Not Access Server field, replication cannot occur.

Database Factors

The next factors that affect replication are at the database layer. This includes the following factors:

- Replica ID and the replica ID cache
- Database ACL
- Replication history
- Replication settings
- Deleted Documents

As soon as another server is granted access for the purpose of replication (i.e., the servers have authenticated and the servers have passed the server access list), the servers check for databases they have in common. The list of databases in common is stored in the replica ID cache. Each server maintains this list of databases that have matching replica IDs. Only the databases in the replica ID cache are replicated.

EXAM TIP
Sometimes databases that are replicas of each other do not replicate. This can occur if the databases are not in the replica ID cache for some reason. Databases may not be in the replica ID cache if the cache has not been updated recently or if replication has been disabled for a database.

NOTE
To enable or disable replication for a database, choose File . . . Tools . . . Server Administration from the menus and select Database Tools. Select the server and database(s) from the list. From the Tool drop-down list, select Replication. Choose Enable or Disable and click Update, as shown in Figure 5.24.

NOTE
The replication for a database can also be disabled in the Replication settings. To verify that replication is enabled, right-click the database icon and choose Replication settings. Select the Other panel. Verify that the option to Temporarily disable replication is not selected, as shown in Figure 5.25.

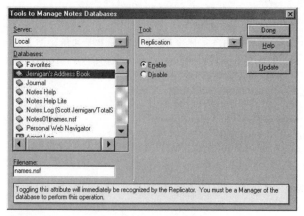

Figure 5.24 Enable or disable replication

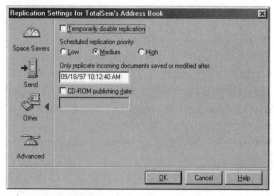

Figure 5.25 Replication settings

After the servers determine which databases can be replicated, the ACL of the database determines the level of replication that can occur. Table 5.4 describes what happens at each ACL level for the servers involved in replication.

> **NOTE**
>
> A server *must* have the same or higher level of Access to a database as the highest user on that server. If the manager of the NAMES.NSF database uses ServerA to open and modify the database, for example, ServerA must have at least Manager access to that database. Otherwise, the changes that that user makes on that server cannot be replicated to other servers. Similarly, if users on ServerA have

Author access to the inventory database, ServerA should have at least Editor access to replicate both newly created and edited documents.

Table 5.4 ACL and Replication

ACL Level	What Can Be Replicated
No access	Nothing can be replicated. If either of the servers has No Access, replication does not continue. Use this option if you need to shut out a particular server at the database level.
Depositor	Do not use Depositor access for a server that needs to replicate. A server with Depositor access cannot replicate changes.
Reader	A server with Reader access to a database can receive changes, but it cannot distribute any changes. This level of access would only be appropriate in a situation where all changes to a database are done at one source location and then distributed.
Author	New documents could be created. Since a server does not author documents, however, this is *not* recommended for an ACL setting for a server.
Editor	New and edited documents can be received and forwarded. This setting is common for any database that will be changed by multiple users at multiple locations.
Designer	In addition to receiving and forwarding new and edited documents, servers with Designer access will receive and forward new and edited design elements. The server that holds the replica of the database that is used by designers to update the design of the database must have Designer access.
Manager	Only one server should have Manager access to a database. Manager access enables the same changes as Designer access with the additional capacity to change the ACL and the replication settings.

Another database level factor that affects whether the database is replicated is the replication history for that database. The *replication history* is a list of dates, times, and servers that have completed replication for a particular database. To access the replication history, shown in Figure 5.26, right-click the database icon on the workspace. When a successful replication is completed, the source server's time is used to issue a time and date stamp on the replication history, and choose Replication history from the shortcut menu. Replication history can be used to aid in troubleshooting replication. If replication has not been occurring as you expected, you can choose the Clear button to remove all replication history entries. This will force a complete replication to occur at the next scheduled replication. Be aware that it will take a significantly longer time, because each element is being verified—not just being compared to the time stamp as usual. If you are only having replication problems with a particular server, you can select that server's replication from the dialog box and click the Zoom . . . button. After you zoom in on a particular event, you can clear only that event.

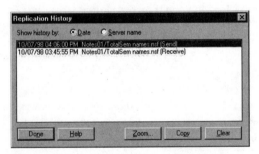

Figure 5.26 Replication history dialog box

 NOTE
Workstation replication does not update the replication history. Only server replication adds this time stamp.

Finally, Replication settings for a database enable control of what can be sent and received during replication. Right-click the database icon to access the Replication settings, as shown in Figure 5.27. The options in the Replication settings dialog box are described in Table 5.5.

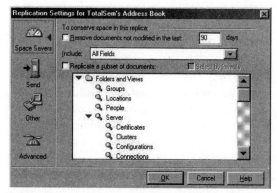

Figure 5.27 Replication settings dialog box

Table 5.5 Replication Settings Options

Panel	Option
Space Savers	This panel enables you to restrict the documents you replicate based on folders, views, or formulas. Most often, laptop users edit these fields to minimize the size of a database. If you choose to restrict the documents you replicate in this dialog box, you are creating a replication formula. In addition, the Remove documents not modified in the last ___ days is an important field for all replicas. This field sets the purge interval for the deletion stubs in the database, as discussed later in the chapter. The purge interval is set to 1/3 the value of this field. The default is 90 days; therefore, the default purge interval is 30 days.
Send	This panel enables you to limit the changes and deletions sent from this replica to other replicas.
Other	This panel allows you to disable replication temporarily if necessary. You also have the option to change the replication priority of the database, as discussed earlier, and to limit incoming replication based on document modified dates.

continues

Table 5.5   Continued

Panel	Option
Advanced	Advanced options enable you to limit incoming documents based on formulas and the specific server (or workstation) sending. This panel allows you to create a complex replication formula that allows the replicas of databases in different locations to contain a different subset of information.

One significant database layer factor affecting replication is how replication deals with deleted documents. When you delete a document from one replica of a database, it should be deleted from *all* replicas of the database. Some users wonder, however, why Notes does not just replace the deleted document with a copy of that document from another replica.

When you delete a document from a database, Notes removes the document but leaves a deleted document identifier, called a *deletion stub*, in its place. This deletion stub is invisible to users, but Notes uses it to know that the document has been deleted. If Notes does not see this deletion stub, it assumes that the document in the other replica is new—and adds it back into the database.

To save space in databases, Notes eventually deletes the deletion stubs. The amount of time that passes before Notes removes the deletion stubs is called the *purge interval*. You can set the purge interval for a database in the Space Savers panel of the Replication settings dialog box, as discussed earlier. The Remove documents not modified in the last ___ day field enables the administrator to set the purge interval. The purge interval is defined as one-third of the value of this field. The default value for this field is 90 days. The default purge interval, therefore, is 30 days.

In some cases, users will complain that deleted documents reappear after replication occurs. If users report this problem, it is likely that the purge interval occurs before the replication interval. To troubleshoot this problem, either increase the purge interval or replicate more often.

Document Factors

In addition to the server and database factors, document layer settings also affect replication. Document layer settings include the following:

- Form and View access lists
- Readers and Authors fields in forms and documents
- Merge replication conflicts

Most settings at the document layer are determined by and are the responsibility of the database designer. Any setting that affects replication, however, needs to be a joint effort between the administrator and designer.

The forms and views in databases have a Security tab on their properties InfoBoxes. These security tabs enable the designer to refine the access given to users by removing them from the Form and View access lists. If you change the default Form and View access, you must include the servers that will be involved in replication.

Similarly, designers can place fields on forms in the database, called Readers and Authors fields, which limit the users who can read and create documents with those forms. If these fields are on forms in databases that will be replicated, they can affect replication at the document layer. The servers that will be replicating the databases must be included in Readers and Authors fields.

Finally, when creating forms in a database, the designer has the ability to enable an option called Merge replication conflicts. Different users may edit the same document on different replicas of a database, which would cause a replication conflict when replication occurs between the databases. In Notes R4.X, there are two features to help prevent replication conflicts. First, Notes uses field-level replication rather than document-level replication. What this means is that Notes only replicates the specific fields that have changed in the documents and databases, rather than replicating the entire document that has changes. Second, Notes gives designers the capacity to enable Merge replication conflicts for the forms in a database. If the Merge replication conflicts option is enabled, as long as users do not change the same field, the changes will be made to the document when replication occurs—and there will be no replication conflict. If, however, both users change the same field, a replication conflict will occur. If the designer did not enable Merge replication conflicts, a replication conflict would occur even if the users edited different fields. To see whether the Merge replication conflicts option has been enabled, open the database to the Design . . . Forms view. Open the Form and access the Form Properties InfoBox by right-clicking the form and choosing Form Properties. The Merge replication conflicts option is on the Basics tab, as shown in Figure 5.28.

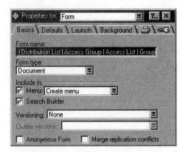

Figure 5.28 Merge replication conflicts option

If both users are editing documents in the same replica of a database, and they edit the same document at the same time, a Save conflict is created. A Save conflict cannot be prevented by enabling the Merge replication conflicts option.

If a Replication or Save conflict is created, it is displayed in the view as a response document, as shown in Figure 5.29. To remove the Replication or Save conflict, open the Replication or Save conflict document. Use the information in the Replication or Save conflict document to determine which changes were not passed on to the main document. Open the main document and make those changes. Save and close the main document. Delete the Replication or Save conflict. This manual merging is the only way to clear a Replication or Save conflict.

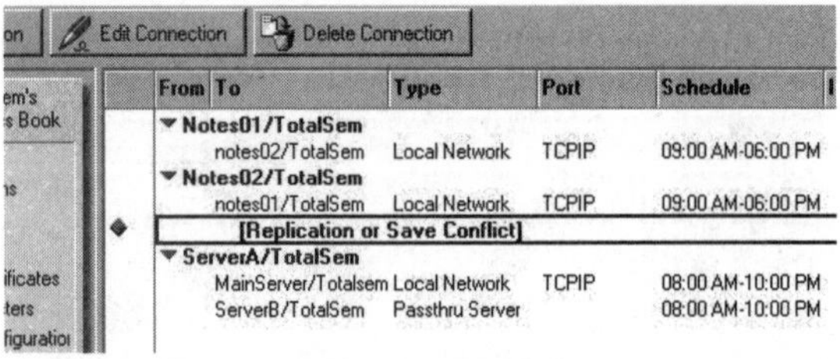

Figure 5.29 Replication conflict shown in a view

In this section, we have discussed what can affect replication at the various levels of a Notes environment. From a security perspective, this gives an administrator more methods of ensuring that only the appropriate data is replicated with the appropriate servers. These replication factors can also be viewed from a troubleshooting perspective. When replication does not occur as you expect, these are some of the first things to examine when troubleshooting the replication.

Process of Replication

It is important to understand the process of replication. Knowledge of the underlying process of replication gives you insight into creating a replication topology and troubleshooting replication. First, we discuss the order in which replication occurs, as well as how this order affects databases. Second, we look at how the replication actually occurs. The better your understanding of the process of replication, the easier replication questions will be on the exam.

Replication occurs in the following order:

1. ACL changes
2. Design changes
3. Document changes

When the ACL of a database has been changed, the ACL replicates first. After the ACL is replicated, the rest of the changes to the databases are made according to the newly updated ACL. The ACL also replicates completely, which means that if changes were made to the ACL on both databases, one set of changes is lost during replication. Let's use the NAMES.NSF database on ServerA and ServerB as an example. The ACL of NAMES.NSF on ServerA, changed 1/9/99 at 2 P.M., contains the following entries:

User	Access
ServerA	Manager
ServerB	Designer
John Doe	Reader

The ACL of NAMES.NSF on ServerB, changed 1/9/99 at 11 A.M., contains the following entries:

User	Access
ServerA	Manager
ServerB	Manager
John Doe	Author

When ServerA contacts ServerB for a pull-push replication, ServerA checks the time stamp and sees that it has the most recent ACL changes. ServerA will not pull the ACL from ServerB, but it will pull any other changes—such as design element and document changes—from ServerB and write those changes to itself. When it begins the push part of the replication, ServerA sees that it has the most recently changed ACL. The ACL of ServerA completely overwrites the ACL of ServerB. When replication finishes, the ACL for both servers looks like the original ACL for ServerA.

Server-to-server replication always begins at the server level, with one server initiating the call to another server based on the Connection document schedule, as described earlier. Until replication is initiated, the server's replicator task is idle. After the call is made, the servers attempt to authenticate, as discussed previously. After authentication occurs, the called server verifies that the initiating server is enabled access by checking its own server document's server access list. Once authentication is complete and access is enabled, the server's replicator task compares the list of databases in the replica ID cache to find the databases available for replication between the servers. After compiling this list, the servers use the replication history to determine the last successful replication. This date is then compared against the last date the database was modified. The database is only replicated if the modification date is more recent than the last replication date.

After determining which databases have been changed, the replicator task builds a list of changes to the database. This list includes ACL, design, and document changes that need to be replicated based on the replication history, the document and field sequence numbers, and the Replication settings. If both databases are Notes r4.X databases, Notes only replicates the changed fields in the documents, rather than the entire documents.

 NOTE
If the changes are made in or replicated to a Notes V3.X database, the entire document is replicated. Field level replication is new to Notes R4.X.

When the replication completes successfully, the source server updates the replication history of the target server with a current time-date stamp.

Workstation-to-Server Replication

Most of the mechanics of workstation-to-server replication are the same as server-to-server replication. One of the main differences is that the workstation software does the work of the replication when replicating with a server—the server's Replicator task does not become active. Another significant difference is the method of initiating and scheduling replication. We discussed earlier the methods of initiating server replication, but here we discuss the methods of initiating workstation replication. The three main methods include the following:

- Initiating replication from the replicator page on the workspace
- Initiating replication from the drop-down list on a stacked icon
- Using a location document to schedule replication

Replicator Page on the Workspace

When you create a replica of a database on your workspace, an entry for it is automatically created on your workspace's replicator page. To examine your replicator page, as shown in Figure 5.30, click the last tab on your Notes workspace, called Replicator. On the replicator page, you see a list of databases that have replicas on another Notes machine.

From the replicator page, you can click on the Start action button to initiate replication for all listed databases that are checked. The checkboxes enable you to select easily the databases that should be replicated. As replication progresses, you will see a status bar at the bottom of the replicator page to let you know what database is currently being replicated and how far in the replication process it has

proceeded. You will also have the ability to skip a database or stop replication, as shown in Figure 5.31.

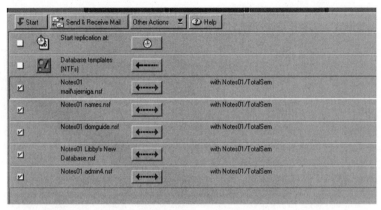

Figure 5.30 Replicator workspace page

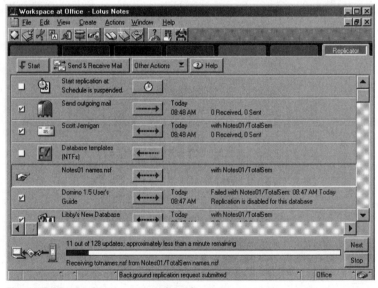

Figure 5.31 Replicator page with replication occurring

You will also be able to obtain the following information from the replicator page:

- The last time that replication occurred for each database
- The server with which replication occurred for each database
- What was replicated at the last replication for each database

The Options button—the blue arrow separating the database name and the last replication time as shown in Figure 5.31—gives you the capacity to specify the server to replicate with and whether to send and/or receive changes. The Options dialog box is shown in Figure 5.32.

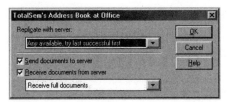

Figure 5.32 Options button

The other action buttons on the replicator page are listed in Table 5.6.

Table 5.6 Other Action Buttons

Action Button	Use
Send and Receive Mail	When users create mail while disconnected, their messages are stored in a Local Outgoing `MAIL.BOX`. Clicking this button enables those messages to be sent—and any waiting messages on the server replica of the mail file to be received.
Other Actions	You have choices to replicate only High-priority databases, replicate only with a particular server, and the two basic options for sending and receiving mail and replicating selected databases.

Stacked Icon Replication

Another option for workstation-to-server replication is to use the replica icons. You can stack your replica icons by choosing View . . . Stack Replica Icons. After the replica icons are stacked, you will see a drop-down arrow on the icons, as shown in Figure 5.33. When you click the drop-down arrow, you can choose which replica to access, and you can choose Replicate to initiate replication.

Figure 5.33 Stacked Replica Icons

Scheduling Workstation-to-Server Replication

Similarly to the way that we scheduled server-to-server replication, you can schedule workstation-to-server replication. To schedule workstation-to-server replication, you will use the Location documents stored in your Personal NAB. Your Location documents can be found in one of two ways. First, you can open your Personal NAB to the Advanced . . . Locations view and either open an existing Location document or create a new Location document. You can also click the Current Location pop-up menu on the status bar in the workspace, as shown in Figure 5.34, and choose Edit Current to edit the current Location document.

Figure 5.34 Current Location pop-up menu

After you have the Location document open, look for the Replication Schedule field and choose Enabled. This command displays the fields for creating the schedule. You can then schedule the replications that are appropriate for that Location. For example, for the Home (modem) location, you may know that you will have your laptop connected every night from 9 P.M. to 11 P.M. You would schedule the replication of the databases you have in common with the server between those times. All databases that have replicas on the selected server will be replicated, based on the databases checked on the Replicator page. You can uncheck databases to disable replication temporarily. The scheduled replications will occur based on the location you have selected at any time. An example of a completed Location document is shown in Figure 5.35.

Long distance prefix:	1		
Area code at this location:			

Mail		Replication	
Mail system:	Notes	Schedule:	Enabled
Mail file location:	Local	Replicate daily between:	09:00 PM - 11:00 PM
Mail file:	mail\sjerniga	Repeat every:	60 minutes
Notes mail domain:	TotalSem	Days of week:	Mon, Tue, Wed, Thu, Fri, Sat, Sun
Recipient name	Personal Address Book Only		

Figure 5.35 Location document completed for scheduled replication

Troubleshooting Replication

Throughout the chapter, we have discussed the topic of replication. You should know by now that setting up and monitoring replication is one of the primary responsibilities of a Notes administrator. Troubleshooting replication is also an important task. As we discussed the replication topics, we mentioned some of the troubleshooting techniques for replication. The following list gives you some suggestions of where to look when you encounter replication problems.

- Are the server names spelled correctly in the Server Connection documents?
- Are there too many replication connections occurring at the same time?
- Are the servers still connected? Is the network or modem not working?
- Is there a replication conflict for the Connection documents?
- Is one of the servers down?
- Does the ACL enable the servers to replicate?
- Can the servers authenticate?
- Is replication disabled on one of the databases?

When you are concerned about replication problems, you should look in your Notes Log (LOG.NSF) for information. The Replication events and Miscellaneous events views are designed to give you information on when and how replication occurs, as well as if and why it does not occur. The Notes log will be discussed in more detail in Chapter 8, "Server Monitoring and Statistics."

Questions

1. Amy created a copy of her Notes database on Server2 from its original home on Server1. When she tries to push the database, no errors are generated—but no changes are replicated. What is a possible way to determine the problem?

   a. Delete the copy and create a new one. The copy may have become corrupt.

   b. Open the database properties of both databases and check the replica ID. They should be the same in order for replication to occur.

   c. Open the access control dialog box for the database and verify that the servers have the necessary rights to replicate changes.

   d. Verify that changes were actually made to the database.

2. Lynette tells you the following items about Notes replication. Which ones are correct?

   a. You must manually start the REPLICA task on the server.

   b. The REPLICA task is automatically started on the server.

   c. If two databases on servers within the same NNN are replicas, they will automatically replicate.

   d. For two databases to replicate, you must force the replication manually.

3. The following is the correct way to create a new replica of a database:

   a. File . . . Database . . . New Copy

   b. File . . . Database . . . New Replica

   c. File . . . Replication . . . New Copy

   d. File . . . Replication . . . New Replica

4. Which server gets new data in a PUSH replication?

   a. The server that initiates the replication

   b. The server that is the destination of the replication

5. Which server gets new data in the PULL replication?

   a. The server that initiates the replication

   b. The server that is the destination of the replication

6. Which is the default type of replication?

   a. Pull-only

   b. Push-only

   c. Pull-push

   d. Pull-pull

7. Which command would force your server (Server1/TotalSem) to call a remote server (Server2/TotalSem) and initiate a replication that would fully synchronize the CERTBOOK.NSF database on both servers?

   a. PUSH Server1/TotalSem CERTBOOK.NSF

   b. PULL Server1/TotalSem Server2/TotalSem CERTBOOK.NSF

   c. REPLICATE Server2/TotalSem CERTBOOK.NSF

   d. REPLICATE Server1/TotalSem CERTBOOK.NSF

8. Which databases are replicated if you do not specify databases in a replication Connection document?

   a. All databases that have replicas on the other server.

   b. No databases are replicated.

   c. NAMES.NSF only

   d. All .NTF files only are replicated.

9. What happens if a connection fails in the middle of a replication when you have the Replication Time Limit set to 15 minutes?

   a. Notes tries to complete the replication every 15 minutes.

   b. Notes tries to complete the replication from the current point —the next time it makes a connection.

   c. Notes will restart the replication in 15 minutes.

   d. Notes will restart the replication at the beginning the next time it makes a connection.

10. If you have a corporate office in Hong Kong where your administrators work and hub offices in Tokyo, London, and New York that must all receive updates and send updates, what type of replication topology should you choose?

    a. Hierarchical topology

    b. Mesh topology

    c. Hub and spoke topology

    d. Wheel topology

11. What is the maximum number of replication requests that can be held in the replication queue?

    a. Two

    b. Three

    c. Four

    d. Five

12. Brian believes that he has correctly configured Passthru for his servers, but replication is not occurring. He created two server Connection documents: one to schedule the replication and one to configure the Passthru. What did he forget?

    a. He needs a third Connection document to define the databases that should use the Passthru server.

    b. He needs a configuration document that enables Server_ Passthru.

    c. He needs to allow the servers to use Passthru by editing the server documents.

    d. He needs to create a Program document to schedule Passthru.

13. Ellen has three Notes r4 servers and two Notes v3 servers in her environment. She wants to configure Passthru replication for her servers. Can she do this in a mixed environment?

    a. Yes. The V3 servers can initiate only the Passthru replication requests.

    b. Yes. The V3 servers can be only the destination of the Passthru replication requests.

    c. Yes. Only the R4 servers can participate in the Passthru replication, however.

    d. No. She must have only R4 servers in her environment to use Passthru for replication.

14. If the Route through field is blank, who can use the server for Passthru?

    a. Everyone

    b. No one

    c. LocalDomainServers

    d. Administrators

15. If the Cause calling field is blank, who can use this server for Passthru?

    a. Everyone

    b. No one

    c. LocalDomainServers

    d. Administrators

16. After two servers have authenticated with each other, what is the next step in replication?

    a. The servers check the ACLs of the databases to be replicated.

    b. The servers check the Replication settings of the databases to be replicated.

    c. The servers check the replica ID cache to make a list of databases they have in common.

    d. The servers check the last replication time and date using the Replication history.

17. What is the minimum ACL level to set for servers that will be replicating a database that has users with Author access?

    a. Author
    b. Editor
    c. Manager
    d. Designer

18. If two servers both have Manager access and one server pulls from the other, what will be replicated?

    a. ACL
    b. Design changes
    c. New documents
    d. Changed documents

19. What replicates first in a standard replication?

    a. ACL
    b. Design changes
    c. New documents
    d. Changed documents

20. What might cause you to clear a replication history, and what will occur?

21. If a user changes the BookName field on a document in the Book Order database, what will be replicated when that database replicates?

    a. All documents
    b. The complete changed document
    c. The changed field on the document
    d. The complete database

22. Aaron and John both edit the BookName field of a document in the Book Order database on different servers. Merge replication conflicts is enabled. What happens when the databases replicate?

    a. The changes are merged into a single document in both databases.

    b. The newer changes are saved and the older changes are discarded.

    c. No changes are made to the document.

    d. A replication conflict is created.

Answers

1. b, c, and d. Open the database properties of both databases and check the replica ID. They should be the same for replication to occur. Open the Access Control dialog box for the database to verify that the servers have the necessary rights to replicate changes. Verify that changes were actually made to the database.

2. b. The REPLICA task is automatically started on the server.

3. d. File . . . Replication . . . New Replica

4. b. The server that is the destination of the replication

5. a. The server that initiates the replication

6. c. Pull-push

7. c. REPLICATE Server2/TotalSem CERTBOOK.NSF

8. a. All databases that have replicas on the other server

9. b. Notes tries to complete the replication from the current point the next time it makes a connection.

10. c. Hub and spoke topology

11. d. Five

12. c. He needs to allow the servers to use Passthru by editing the server documents.

13. b. Yes. The V3 servers can be only the destination of the Passthru replication requests.

14. b. No one

15. b. No one

16. c. The servers check the replica ID cache to make a list of databases they have in common.

17. b. Editor. Author access would enable the server to replicate any new documents, but not any documents that have been edited. Users with Author access can edit their own documents.

18. All of the above that have been changed, starting with the ACL.

19. a. ACL

20. You may choose to clear a replication history if some changed documents or other elements are not replicating between two servers. If you clear the replication history, a complete replication of all elements will occur.

21. c. The changed field on the document; Notes R4.X uses field-level replication.

22. d. A replication conflict is created because the changes were both made to the same field.

CHAPTER 6

Administration Tools and Tasks

A s a Notes System Administrator, you are expected to maintain the environment that you have created. In previous chapters, we planned and configured your Notes environment, registered and installed users and servers, created replication and mail routing schedules, and implemented security. The focus of this chapter is on maintaining and administering the elements that you have already created. We describe the use of a new Notes R4.X process, called the Administration Process, to rename and recertify users and servers. We also describe the use of the server console and the server administration panel, including some of the more useful console commands. We discuss the ability to assist end users with configuration and communication through Profile documents and Remote Passthru. Finally, we discuss the need for backing up your Notes data.

Objectives

After reading this chapter, you should be able to answer questions based on the following objectives:

- Rename and recertify users and servers.

- Rename and recertify users and servers using the Administration Process.

- Use the Administration Process to update documents in the NAB.
- Use the Server Administration Panel to control the server.
- Create database quotas and thresholds using the Server Administration panel.
- Control the server from the server console.
- Use Profile documents to expedite user configuration.
- Assist users with connection to servers using Remote Passthru.
- Troubleshoot user connections using network connection tracing.
- Back up and verify Notes data.

The Administration Process

System administrators spend a good deal of time and energy renaming and recertifying users. In a normal Notes environment, elements change constantly. Users change their names and move between organizational units. New organizational units are created for users or servers. In addition, you may create or delete groups. Users may leave and need to be removed from the NAB. Although all of these processes are important for administrators, the two most complex tasks are renaming and recertifying users. This section will concentrate on those processes.

All of these processes, including renaming and recertifying users, can be accomplished manually or automatically in Notes R4.X. To rename and recertify users manually, use the Rename and Recertify Person actions available in the NAB. To rename or recertify a user automatically, use the Administration Process new to Notes R4.X. This section describes both methods in detail.

Renaming a User

To change a user's common name, open the NAB to the People view and select the Person document of the user who you need to rename. Choose Actions . . . Rename Person from the menus. The Certify Selected Entries dialog box is displayed, as shown in Figure 6.1. Choose the Change Common Name button.

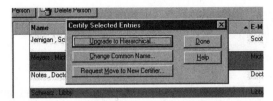

Figure 6.1 Certify Selected Entries dialog box

You will be prompted to select the correct certifier, as shown in Figure 6.2, and in this case it should be the same certifier with which the ID was originally created. After selecting the correct certifier ID, you will be prompted for the password. Type in the password.

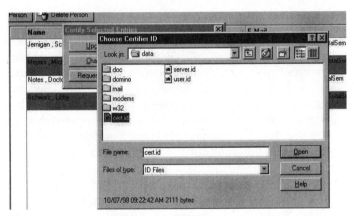

Figure 6.2 Choose the correct certifier.

The Rename Selected User dialog box is displayed, as shown in Figure 6.3. Make the necessary changes to the user's common name and click the Rename button.

After renaming the selected user in the NAB, the user's name must be manually updated in any database ACLs, as well as in any Reader and Author name fields.

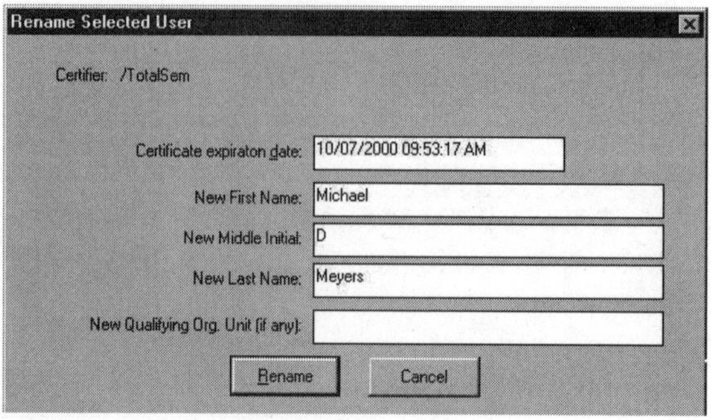

Figure 6.3 Rename Selected User dialog box

Recertifying a User or Server

A user or server may move within the organizational structure. If this occurs, the user may need to be recertified under a different organizational unit. Users and servers will also need to be recertified when their certificates expire. Certificates for users expire after two years, while certificates for servers and certifiers expire after 100 years.

To recertify a user, open the NAB to the People view and select the Person document of the user who you need to recertify. Select Actions . . . Recertify User from the menus. The Choose Certifier ID dialog box is displayed. Choose either the same certifier (if you are renewing the certification) or a new certifier (if you are moving the user). After typing in the correct password, the Renew Certificates in Selected Entries dialog box is displayed, as shown in Figure 6.4. Change the certification expiration date, if necessary, and click Certify.

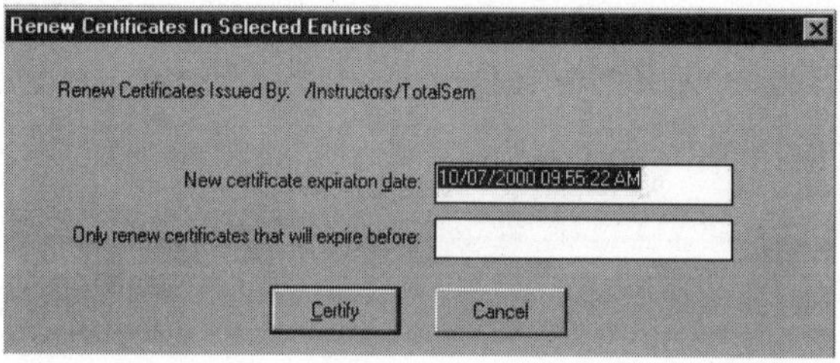

Figure 6.4 Renew Certificates In Selected Entries dialog box

Again, you would make any changes to ACLs, Form and View access lists, and Reader and Author fields manually.

To recertify a server, the steps are similar. Open the NAB to the Server . . . Servers view, and select the Server document for the server that you are recertifying. Select Actions . . . Recertify Server from the menus. The remaining steps are the same.

As an administrator, one of the ways that you will receive requests to recertify or rename users is through your mail file. You may remember that when you created certifiers, there was a field for Administrator. This field enabled you to enter the name of the administrator who was responsible for that certifier. Users can then request to be renamed using the following steps.

To request a name change, choose File . . . Tools . . . User ID from the menus. In the User ID dialog box, select the More Options panel, as shown in Figure 6.5.

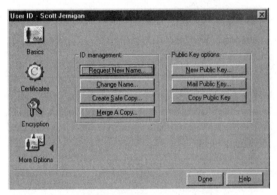

Figure 6.5 User ID dialog box

Choose the Request New Name button. This button prompts you to enter the name that you would like to use, as shown in Figure 6.6. When you choose OK in the dialog box, you are prompted to the Mail New Name Request dialog box, shown in Figure 6.7. Select a name, or type the name of an administrator, and click Send. If an administrator's name was added to the certifier, it will appear in this field by default. Notes prompts you to confirm the name that you want to change and then sends the request. The name change request is delivered to the administrator's mail file.

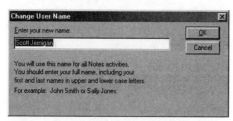

Figure 6.6 Change User Name dialog box

Figure 6.7 Mail New Name Request dialog box

After you receive a name change request in your mail file, the following steps complete the renaming. Open the message. The default subject of the message is: "My ID containing a new name is attached. Please certify it and send it back to me by using the Actions menu Certify Attached ID File . . . option." This subject line tells you precisely what to do with the message. With the message open, choose Actions . . . Certify Attached ID file . . . from the menus. You will be prompted to choose the correct certifier. Select the certifier that was used to create the user and type the password. The Certify ID dialog box, shown in Figure 6.8, is displayed. Choose the appropriate registration server and make any other necessary changes to the certifier expiration date or to the minimum password length and then choose Certify. After certifying the ID and updating the NAB, Notes displays the Mail Certified ID dialog box, shown in Figure 6.9, to enable you to send the recertified ID file back to the user. After receiving the recertified ID, the user needs to choose Actions . . . Accept Certificate to complete the process.

Again, after renaming or recertifying any user, the administrator must manually change any ACL entries for that user, including those for the user's mail file.

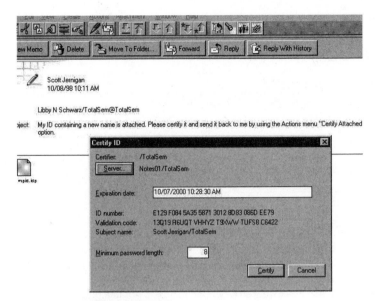

Figure 6.8 Certify ID dialog box

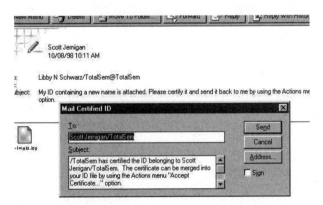

Figure 6.9 Mail Certified ID dialog box

Using the Administration Process

Although it is important for administrators to know how to perform renaming and recertifying manually, Lotus added a new server
feature to Notes R4.X to automate the process. This feature is called

the Administration Process, or AdminP. AdminP can automate a variety of Notes administration tasks, including the following jobs:

- Renaming users
- Recertifying users and servers
- Deleting users, servers, and groups
- Converting users and servers from flat to hierarchical naming
- Creating and deleting mail files
- Adding resources to and deleting them from the NAB when these actions are initiated in the Resource Reservations database
- Creating replicas of multiple databases
- Enabling password checking during authentication
- Adding and removing servers from clusters

> **NOTE**
> These processes can be completed using AdminP only when users and servers in the organization use hierarchical naming. The only exception to this rule is the process of upgrading users from flat naming to hierarchical naming.

When you configure and start a Notes R4.X server, the AdminP runs as a server task. You can start and stop the AdminP task from the server console by typing the following commands:

```
TELL ADMINP QUIT and LOAD ADMINP
```

In addition, the AdminP task is automatically listed in the SERVERTASKS= line in the NOTES.INI, which means that it will start automatically at server startup.

Other tasks and minimum requirements, however, must be completed before the AdminP can be used. These tasks and requirements are described next.

> **EXAM TIP**
> AdminP is one of the key elements for the System Administration I exam. Main topics that are covered on the exam focus on the prerequisites for using AdminP, what AdminP can be used to accomplish, and what advantages are gained by using AdminP. These items are described in this section. In addition, some of the details presented here will help you understand and accomplish the process—but are not necessarily tested on the exam.

ADMINP PREREQUISITES

As mentioned earlier, the first prerequisite to use the AdminP is hierarchical naming in your organization. If you are not using hierarchical naming, AdminP can only be used to convert to hierarchical naming.

The following items are the other prerequisites for using AdminP in your environment:

- *All servers involved in the changes must be running Notes R4 or higher.* The AdminP is new to Notes R4.X and is not reverse-compatible, in that it will work only on servers that have Notes R4.X installed. If you have servers that use previous versions of Notes, changes cannot be made directly on those servers using AdminP. The changes will either have to be replicated from other servers or made manually.

- *ADMIN4.NSF, the Administration Requests database, must exist.* When a Notes R4.X server is installed, the ADMIN4.NSF database is automatically created. The Administration requests database enables administrators to observe and track the activities being completed by the AdminP. Any documents created when Rename or Recertify requests are started are stored in this database.

- *To perform the administration requests correctly, the administrator who uses AdminP or initiates rename, recertify, or delete processes must have the correct level of access in both the Administration requests database and in the NAB.* The user making the changes in the environment must have at least Editor access in the ACL of the ADMIN4.NSF database. The LocalDomainServers will require Manager access in the ACL to replicate changes correctly.

- *A server must be designated as the Administration Server of the NAB.* One server, and one server only, should have this designation. This server will receive the requests to rename, recertify, delete, and change users, servers, and groups. If you have more than one server designated as the Administration Server of the NAB, you may receive unpredictable results—such as Replication and Save conflicts or other errors. To designate a server as the Administration Server for the NAB, right-click the icon for the NAB and select Access Control. Choose the Advanced panel, as shown in Figure 6.10. Select an appropriate server from the drop-down list.

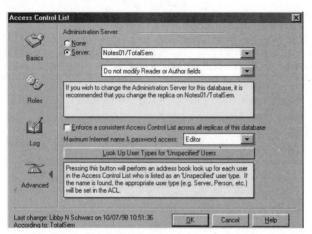

Figure 6.10 Designate an Administration Server for the NAB

When you make this change to the NAB, it is best to stop the server by typing Q or E at the server console before making the change. In addition, if the server has been running without this setting in the NAB, you will want to delete the current ADMIN4.NSF and enable the server to create a new one when you restart it. Remember to delete the ADMIN4.NSF database from the other servers as well, because they are all replicas of the one on the first server.

- CERTLOG.NSF, *the Certification Log database, must exist on the server designated as the Administration Server of the NAB.* Administrators must have at least Author access in the ACL. CERTLOG.NSF maintains a record of all IDs certified in your Notes organization. This database must exist for the AdminP to work correctly. In addition, administrators should have at least Author access to the database, so when they certify users, servers, and organizational units, the record of this new ID can be added to the Certification Log.

- *Any databases you want to be affected by the AdminP must have an administration Server designated.* You must designate an Administration server for all databases that you want to be affected by the AdminP, which is similar to the need for designating an Administration Server for the NAB. If you rename a user, for example, and you want the AdminP to rename that user in the ACL of the databases on your server, you must indicate an Administration Server for those databases. If there is no Administration Server designated in the Advanced Panel of the ACL dialog box for a particular database, the AdminP will skip that database. Renamed or recertified users may lose access to the database when this occurs.

You can designate Administration Servers for databases individually, as described earlier, or you can designate the Administration Server for a group of databases at one time. To set this option for multiple databases, choose File . . . Tools . . . Server Administration from the menus. In the Server Administration panel, choose the Database tools button. The Tools to Manage Notes Databases dialog box, shown in Figure 6.11, is displayed. Select the databases for which you want to set this option. You can select multiple databases by pressing *Shift* or *Ctrl* when choosing databases.

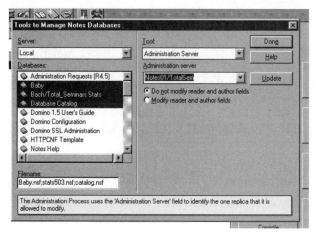

Figure 6.11   Tools to Manage Notes Databases dialog box

After selecting the databases, choose Administration Server from the Tool drop-down list. Choose the correct Administration Server. Choose whether the Administration process should modify Reader and Author fields. Then click Update.

To verify any particular database's Administration Server, open the ACL dialog box. In the ACL list, the server with a key as part of the icon, as shown in Figure 6.12, is the designated Administration Server.

Figure 6.12 Verifying the Administration Server for a database

CONFIGURING AND USING THE ADMINP

After you have completed all the prerequisites for using AdminP in your environment, you can begin using the AdminP to complete administration tasks. You will find it helpful, however, to know some of the default configurations and to be able to change those as necessary.

Each of the types of requests that the AdminP can perform has a priority. These priorities are set in the Administration Process section of the Server document, as shown in Figure 6.13. The following list describes these priorities:

- Immediate requests are completed every minute.

- Interval requests are completed at the interval specified in the Server document, or every 60 minutes.

- Daily requests are completed once a day at the time specified in the Server document, or at midnight.

- Delayed requests are completed at the day and time specified in the Server document, or at midnight on Sundays.

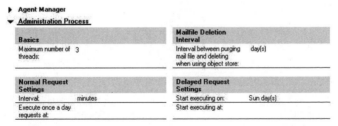

Figure 6.13 Administration Process options in the Server document

Some of these priorities can also be set using server configuration documents or lines in the NOTES.INI file.

Each of the tasks that AdminP performs is predefined to take place at one of the intervals described earlier. Table 6.1 shows some examples of these request types and their intervals. For additional information, refer to the Notes Administration Help database or the Notes *Administrator's Guide* available from Lotus.

Table 6.1 Administration Request Intervals

Request Type	Priority Type
Initiate Rename in Address Book	Interval
Rename Person in Address Book	Interval
Rename in Reader/Author fields	Delayed
Rename in Person documents	Daily request
Delete in Address Book	Interval
Delete in Person documents	Daily request
Delete in Access Control Lists	Interval
Request File Deletion	Immediate

In addition, you can use console commands to force the AdminP to execute by typing the following command at the server console:

```
TELL ADMINP PROCESS REQUEST
```

The request can be one of the following elements:

- INTERVAL
- DAILY
- DELAYED
- ALL

These options force AdminP to process the requests that use the specified type of priority. If you have a Delete in Address Book request pending, for example, you might type either TELL ADMINP PROCESS INTERVAL or TELL ADMINP PROCESS ALL to force action on the request.

You can also use the server console commands to verify which databases have the Administration Server set. If you type the command TELL ADMINP SHOW DATABASES, all databases that have the current server set as their Administration Server will be listed.

HOW ADMINP WORKS

As mentioned earlier, you can use the AdminP to automate the process of renaming or recertifying users, along with other tasks. In the earlier sections, we described the prerequisites, benefits, and options of AdminP. In this section, we will explain the steps that would occur in a Rename User request that uses the AdminP. We will give only one example, but you should have an idea of what occurs.

First, be aware that both the administrator and the user still have tasks to complete in this process. The user must still request a name change and specify the new name. The administrator must still choose to rename the user, using either the action in the NAB or the action available if the user requested the name change via Notes mail. After this point, the AdminP takes over and completes the following steps:

1. An Initiate Rename in Address Book document is created in the Administration Requests database on the server initiating the request. This document is replicated from the initiating server to the Administration Server of the NAB and to any other servers, according to the replication schedule.

2. The Certification Log on the server initiating the request is updated with the new information. The Certification Log on the Administration Server and on any other servers is also updated, according to the replication schedule.

3. After it receives the request and reaches the specified interval, the Administration Server for the NAB updates the Person document in the NAB. When the Person document is updated, the status is posted in the Administration Requests database.

4. The NAB is replicated to all other servers in the domain, based on the replication schedule.

5. When the user tries to authenticate with any server in the domain that has the updated information, the user is prompted to accept the name change. After the user accepts the name change, the workstation updates the user's ID file with the new certificate.

6. A Rename Person in Address Book document is created in the Administration Requests database on the server with which the user authenticated. This document is replicated from the initiating server to the Administration Server of the NAB and to any other servers, according to the replication schedule.

7. When the Administration Server of the NAB receives the request and reaches the specified interval, it updates all instances of the user's name in the NAB, including groups, Server documents, and Connection documents.

8. A Rename in Person Documents document is created in the Administration Requests database on the Administration Server of the NAB. The Person Document is updated at the time set for this request in the Server document.

9. A Rename in Access Control List document is created in the Administration Requests database on the Administration Server of the NAB. This document is replicated from the Administration Server of the NAB to any other servers, according to the replication schedule.

10. After each server receives this request and reaches the specified interval, the AdminP updates the name in the ACL. The update occurs for the databases stored on that server that have that server's name listed as the Administration Server.

As you can tell, the AdminP is a time-intensive process based on replication schedules and AdminP intervals and priorities. If you have an urgent request, you could speed up the process with the following steps:

- Initiate the request on the Administration Server of the NAB.

- Use console commands to force AdminP to process requests.

- Force the Administration Requests database and the NAB to replicate ahead of their replication schedules.

Although the process sounds long and unwieldy when written out, it saves the administrator from finding and updating all the groups, ACLs, Reader and Author name fields, Connection document fields, and other fields manually. The portion of the process that requires an administrator's intervention ends after configuring the process once and then initiating the request. Other processes that use AdminP require similar steps.

Controlling and Maintaining the Server

As a system administrator, you are expected to maintain and control activities on the server. You can do this activity in three main places:

- The Public NAB
- The server console
- The Server Administration panel

We have already discussed and described the NAB and its uses in Chapter 2, "The Notes Name and Address Book." This section will focus on the uses of both the server console and the Server Administration Panel.

SERVER CONSOLE

The server console is a command prompt-based interface that exists on the Notes server machine. Also called the server shell, the server console must be running for the server to be considered running. When you exit or close the server console, you are bringing down your Notes server. You can use the server console to type in commands that will return values either directly to the screen, to a text file, or to the Notes Log file. The commands can either perform an action or obtain information.

EXAM TIP
The following console commands are not necessarily all tested on the exam, but it is useful to know a variety of commands because they will be represented on the exam. You should know how to input the commands with the correct syntax, as well as know what the command will return.

The following commands are some of the most common commands used at the Notes server console. Note that most commands also show an abbreviation. The commands in Notes can be abbreviated to the shortest length that will still enable them to be unique.

- SHOW SERVER or SH SE shows basic server information, including the name and server directory. This command can be used to view dead and pending mail, server uptime, transactions, and the Shared mail status and database. The results of a SHOW SERVER command are displayed in Figure 6.14.

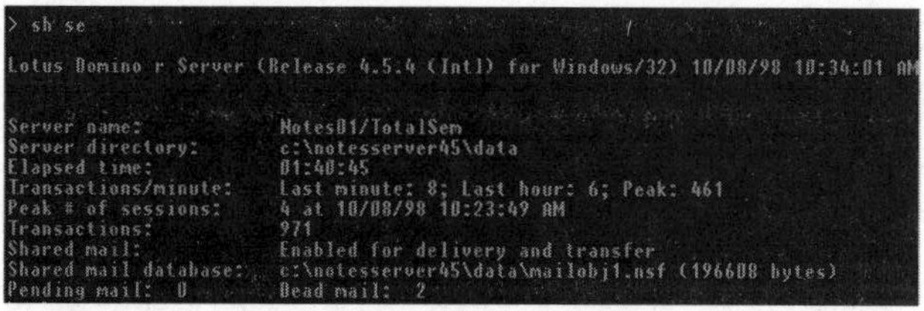

```
> sh se
Lotus Domino r Server (Release 4.5.4 (Intl) for Windows/32) 10/08/98 10:34:01 AM

Server name:              Notes01/TotalSem
Server directory:         c:\notesserver45\data
Elapsed time:             01:40:45
Transactions/minute:      Last minute: 8; Last hour: 6; Peak: 461
Peak # of sessions:       4 at 10/08/98 10:23:49 AM
Transactions:             971
Shared mail:              Enabled for delivery and transfer
Shared mail database:     c:\notesserver45\data\mailobj1.nsf (196608 bytes)
Pending mail:    0        Dead mail:  2
```

Figure 6.14 SHOW SERVER command

- SHOW TASKS or SH TA shows the basic server information available with SHOW SERVER, plus a list of the tasks currently running on the server. This list includes some information about active users, as well as all the idle and active tasks. The results of a SHOW TASKS command may be long, as shown in Figure 6.15, and therefore, SH SE may be more useful in many situations.

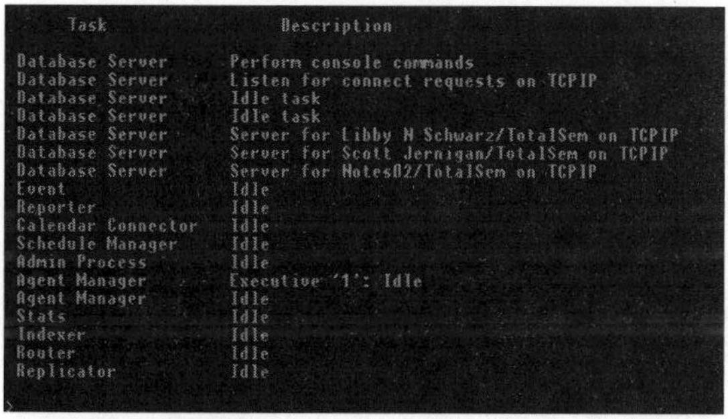

```
    Task                 Description
Database Server      Perform console commands
Database Server      Listen for connect requests on TCPIP
Database Server      Idle task
Database Server      Idle task
Database Server      Server for Libby N Schwarz/TotalSem on TCPIP
Database Server      Server for Scott Jernigan/TotalSem on TCPIP
Database Server      Server for Notes02/TotalSem on TCPIP
Event                Idle
Reporter             Idle
Calendar Connector   Idle
Schedule Manager     Idle
Admin Process        Idle
Agent Manager        Executive '1': Idle
Agent Manager        Idle
Stats                Idle
Indexer              Idle
Router               Idle
Replicator           Idle
>
```

Figure 6.15 SHOW TASKS command

- SHOW USERS or SH US displays a list of active users. The list includes the name of the user or server, the databases they are using, and the number of minutes since they last used those databases. An example of a SHOW USERS command is shown in Figure 6.16.

```
> sh us
  User Name            Databases Open       Minutes Since Last Used

Libby N Schwarz/TotalSem
                       mail\lschwarz.nsf              2
Scott Jernigan/TotalSem
                       names.nsf                      1
Notes02/TotalSem
```

Figure 6.16 SHOW USERS command

- BROADCAST "Message" "UserName" enables an administrator to send a message to a user who is currently connected to the server. Use a SHOW USER command to verify which users are connected, and then send them a message using BROADCAST. The user name should be hierarchical and enclosed in quotes. An example of this command might be: BROADCAST "Server going down in five minutes." "Doctor Notes/ TotalSem." If you do not specify a name, the message is sent to all currently connected users. The message appears to the user in the workstation status bar, as shown in Figure 6.17.

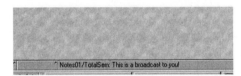

Notes01/TotalSem: This is a broadcast to you!

Figure 6.17 Result of a BROADCAST command

- DROP ALL enables you to disconnect all users, while DROP "UserName" enables you to disconnect a currently connected user. This command closes their connection to the server. The result of a DROP ALL command is shown in Figure 6.18.

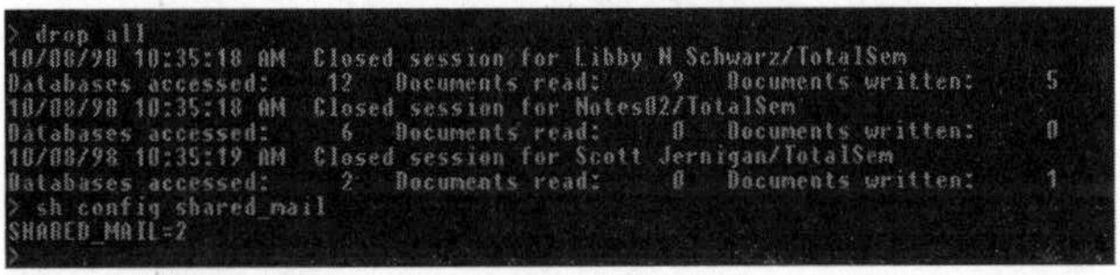

```
> drop all
10/08/98 10:35:18 AM  Closed session for Libby N Schwarz/TotalSem
Databases accessed:    12   Documents read:    9   Documents written:   5
10/08/98 10:35:18 AM  Closed session for Notes02/TotalSem
Databases accessed:     6   Documents read:    0   Documents written:   0
10/08/98 10:35:19 AM  Closed session for Scott Jernigan/TotalSem
Databases accessed:     2   Documents read:    0   Documents written:   1
> sh config shared_mail
SHARED_MAIL=2
```

Figure 6.18 DROP ALL command

- SHOW CONFIG Variable enables the administrator to view the value of a variable that is set in the NOTES.INI. SHOW CONFIG SHARED_MAIL, for example, displays the value of the SHARED_MAIL variable. Another example would be SHOW CONFIG REPLICATORS, which would display the number of replicators enabled on the server. To see a complete list of available settings, refer to the Notes Administration Help database. An example of a SHOW CONFIG command is shown in Figure 6.19.

```
> sh config servertasks
SERVERTASKS=Replica,Router,Update,Stats,AMgr,Adminp,Sched,CalConn,Report,Event
> sh config shared_mail
SHARED_MAIL=2
> sh config replicators
REPLICATORS=3
>
```

Figure 6.19 SHOW CONFIG command

■ SET CONFIG Variable enables the administrator to set the value of a variable that is stored in the NOTES.INI. SET CONFIG SHARED_MAIL=1, for example, changes the value of the SHARED_MAIL variable. Another example would be SET CONFIG REPLICATORS=2, which would enable two replicators on the server. To see a complete list of available settings, refer to the Notes Administration Help database.

■ TELL enables the administrator to initiate a specific Notes task, such as TELL REPLICA QUIT or TELL ROUTER QUIT.

■ LOAD enables the administrator to load a specific Notes task, such as LOAD REPLICA or LOAD ROUTER.

■ ROUTE Servername enables the administrator to force mail routing to the specified server immediately, rather than waiting for the schedule specified in the Connection document. For more information about mail routing, refer to Chapter 4, "Notes Mail."

■ REPLICATE Servername Databasename enables the administrator to force a complete Pull-push replication with the specified server. If the databasename parameter is also used, only the specified database will be replicated. Otherwise, all databases with replicas in the \NOTES\DATA directory and subdirectories will be replicated. For more information about replication, refer to Chapter 5, "Notes Replication."

■ PUSH Servername Databasename enables the administrator to force a Push-only replication with the specified server. If the databasename parameter is also used, only the specified database will be replicated. Otherwise, all databases with replicas in the \NOTES\DATA directory and subdirectories will be replicated. For more information about replication, refer to Chapter 5, "Notes Replication."

■ PULL Servername Databasename enables the administrator to force a Pull-only replication with the specified server. If the databasename parameter is also used, only the specified database will be replicated. Otherwise, all databases with replicas in the \NOTES\DATA directory and subdirectories will be replicated. For more information about replication, refer to Chapter 5, "Notes Replication."

- SET SECURE Password enables the administrator to secure the console itself with a password. Users or other administrators will not be able to type in console commands without the password if this has been set.

- HELP gives the administrator an overview of the available console commands, including use and syntax.

- QUIT or EXIT enables the administrator to exit the server console. This command shuts down all server tasks prior to shutting down the server.

Other examples of server commands include Show Directory, Show Diskspace, Show Statistics, Show Memory, Show Performance, Show Schedule, and Show Port. For more information about these and other server commands, refer to the Notes Administration Help database.

Server Administration Panel

Another method of controlling the Notes server and environment is through the Server Administration panel. The Server Administration panel can be accessed through a Notes workstation by choosing File . . . Tools . . . Server Administration from the menus. This command displays the panel, as shown in Figure 6.20.

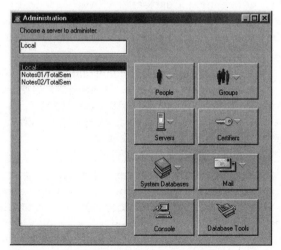

Figure 6.20   Server Administration panel

The Server Administration panel gives an administrator the capacity to do the following activities:

- Register users, servers, organizations, and organizational units. For more information on users, servers, organizations, and organizational units, see Chapter 1, "Installation and Configuration."

- Create groups. See Chapter 1, "Installation and Configuration," and Chapter 2, "The Notes Name and Address Book," for more information on groups.

- Access the NAB for information on users, servers, groups, organizations, and organizational units. See Chapter 1, "Installation and Configuration," and Chapter 2, "The Notes Name and Address Book," for more information.

- Send console commands to servers.

- Use database tools to administer databases on Notes servers.

- Configure servers.

- Administer Notes system databases, such as the Log, and statistics reporting. For more information on system databases in Notes, refer to Chapter 8, "Server Monitoring and Statistics."

- Send a mail trace. For more information on using Mail Trace, see Chapter 4, "Notes Mail."

- Access the server's MAIL.BOX. For more information on using MAIL.BOX, see Chapter 4, "Notes Mail."

This section describes the use of the Server Administration panel for sending console commands to the server and administering databases.

REMOTE CONSOLE

In the earlier section, we discussed an administrator's ability to control the server using the server console. For administrators who do not have direct access to the server console for a machine, or for administrators who prefer to administer servers remotely, Notes included a remote console accessible in the Server Administration panel. All of the console commands available at the server console are also available at the Remote console.

After opening the Server Administration panel, select the server to which you want to send commands by clicking the server's name in the list. If the server is not in the list (e.g., a server in a different NNN), you can access the server by typing in the hierarchical name of the server. Then click the Console button. The Remote Console dialog box is displayed, as shown in Figure 6.21.

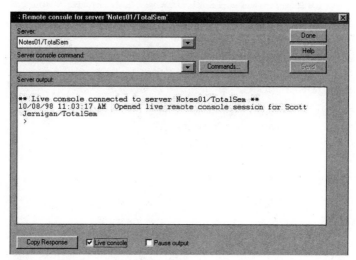

Figure 6.21 Remote Server Console

To see the commands act on the server immediately, and to see other server activity as it occurs, choose the Live Console check box. To see only the results of your commands, choose Pause Output. You can type or select a command in the Server Console command field. To select a command, click the Commands... button. This displays a list of available commands, including a description, as shown in Figure 6.22.

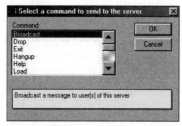

Figure 6.22 Choose a Console Command.

After typing or selecting a command, press ENTER or click the Send button. This sends the command to the server console. If you can see the actual server console while using the remote console, you will notice that the remote console commands are displayed on the server after you send them.

To copy the response from the server so that you can paste it into another document or application, click the Copy Response button.

One of the benefits of the remote console is that it gives you the capacity to scroll through your remote console session. This feature is especially useful when typing in a command that returns a long list of values, such as SHOW TASKS or SHOW USERS. In addition, you can scroll through the list of commands that you sent in a particular remote console session by clicking the drop-down arrow next to the Server console commands field.

 NOTE
To use the Remote console, your name must be listed in the Administrator's field of the Server document in the NAB.

DATABASE TOOLS

In the Server Administration panel, the administrator has a variety of tools to control and administer databases. To access these tools, click the Databases button (Database Tools after Notes R4.5) on the Server Administration panel. Some of the tools available include the following items:

- *Administration Server*, which has the capacity to set an Administration Server for a database or group of databases, as discussed earlier in this chapter.

- *Compact*, which enables the administrator to remove unused white space from databases. This feature reduces the overall size of the database on disk and makes it more efficient.

- *Consistent ACL*, which provides the administrator with the capacity to enable or disable a Consistent ACL across all replicas of a database. This feature is especially useful in environments that use the AdminP.

- *DB Fixup*, which enables the administrator to search a database for corruption and inconsistencies. This process deletes the items that are corrupt so that they can be replicated into the database from another replica.

- *Quotas*, which enable an administrator to set a database quota and a warning threshold on a database. The following section describes this option in more detail.

In Notes R4.X, the maximum, hard-coded size Limit for a database is 4GB. The default Limit when creating a new database or a new replica is 1GB. Users will not be able to save documents in the database once it reaches its database size Limit. In addition, the database Limit cannot be changed after it is set. The only way to change a database Limit is by creating a new database. To set a Limit when creating a database, use the Size Limit button on the New Database dialog box, as shown in Figure 6.23.

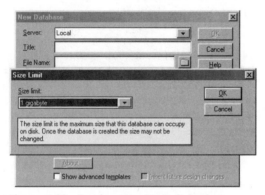

Figure 6.23   Database Size Limit dialog box

Because this Limit is absolute, it is to an administrator's advantage to have some warning when the database is approaching this Limit. To create a warning, Lotus enables an administrator to set two types of soft boundaries on the size of databases, called Quotas and Warning Thresholds.

A Quota, set using the database tools shown earlier, is a restriction on the size of the database. If the database reaches the Quota, users will not be able to save documents in the database and will receive an error. Unlike the hard-coded Limit, however, this Quota can be changed in the same way that it is set, as shown in Figure 6.24.

The reason for setting a database Quota is so that you as the administrator will receive information about the size of the database as it approaches its hard-coded Limit. Set the Quota near, but obviously under, the hard-coded Limit.

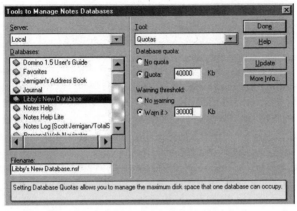

Figure 6.24   Set a database Quota and Warning Threshold.

A Warning Threshold is an administrator's other opportunity to receive information about the size of a database. If a Warning Threshold is set using the database tools, Notes will place a warning message in the Notes Log when the database reaches this threshold. No error will be displayed to the user when the database is over its Warning Threshold. For this reason, it is usually recommended to set a warning threshold slightly lower than the Quota, to prevent users from receiving errors when they try to save documents.

 NOTE
Other database administration tools are described in Chapter 8, "Server Monitoring and Statistics." Additional database administration tasks, such as running FIXUP and UPDALL, and backing up your server, are described at the end of this chapter.

OTHER ADMINISTRATION TOOLS AND TASKS

The Server Administration panel also offers other tools to accomplish administration and troubleshooting tasks. Some of these additional options are described next.

When a warning or an error occurs in Notes, it will usually display the error on the Notes server console. If you are like most administrators, however, you will not be sitting by your server waiting for errors. All activity that occurs on the server is therefore also written to the Notes Log database (LOG.NSF). You can view the Log by choosing File . . . Database . . . Open from the menus and accessing any server. Select the Notes Log database for the appropriate server from the database list.

You can also view the Log using the Server Administration panel. Open the Server Administration panel and choose the server whose Log you want to view. Then click the System Databases button and select Open Log from the list.

The Log shows all types of events that occur on the server, including replication and mail routing events and errors. This feature is one of the best ways to troubleshoot or verify the health of a server. An administrator should open and check the Log multiple times daily. The Miscellaneous Event view of the Log enables you to open Log documents that show all activity on the server, as shown in Figure 6.25. For more information on using and analyzing a server Log, refer to Chapter 8, "Server Monitoring and Statistics."

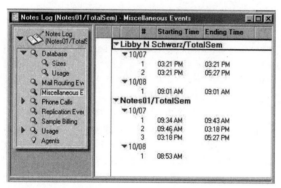

Figure 6.25   Notes Server Log

End User Administration Tasks

Administrators have a great deal of responsibility for assisting users in their environment. Some of the ways they assist users have already been discussed, such as troubleshooting mail, security, and replication. Creating users is also an administrative task directly related to end users. Users will also need assistance setting up their workstations and connecting to the server. In Chapter 1, "Installation and Configuration," we discussed the basics of creating users and setting up workstations. In this section, we will describe some additional end user administration options. These options are User Profiles, Remote User Passthru, and troubleshooting and tracing network connections.

USER PROFILES

When you registered a user in Chapter 1, "Installation and Configuration," you may have noticed the field on the Register Person dialog box called Profile. The User profile enables an administrator to set a variety of workstation defaults for users, including Internet settings, Passthru settings, and the databases that appear on their workspaces. To create a User profile, open the NAB to the Server . . . Setup Profiles view. Click the Add Setup Profile button to display the New Setup Profile form, shown in Figure 6.26.

USER SETUP PROFILE

Basics	
Profile name:	Accounting
Internet browser:	Notes
Retrieve/open pages:	from Notes workstation

Default Databases	
Database links:	

Default Passthru Server	
Server name:	MainServer
Country code:	
Area code:	214
Phone number:	555-2233

Default Connections to Other Remote Servers			
Server Names	Country Codes	Area Codes	Phone Number
ServerC		409	555-2122

Figure 6.26   User Setup Profile

1. First, type in a name for the profile. This name should be descriptive of the users for whom the profile will be used.

2. Complete information describing the user's Internet and Passthru defaults.

3. Complete any other necessary information and then save the document.

EXAM TIP
Your ability to create and use User Setup Profiles is tested on the exam, as well as your knowledge of the situations in which a User Setup Profile can be used. The specifics of the documents, however, are not tested.

If User Profile documents are in the NAB when you begin registering a user, they are available in the drop-down Profiles list in the Registration dialog box.

REMOTE PASSTHRU USE

Passthru is a new tool for Notes R4.X that enables users to connect to multiple servers through a single dial-up or Internet connection. In Chapter 5, "Notes Replication," we discussed Passthru as it relates to servers and replication. In this chapter, we discuss Passthru primarily as it relates to remote, dial-up connections made by users.

In the past, when users dialed in to their Notes servers, they had to make a separate phone call for each server with databases that they needed. With Passthru, however, users can make one call and connect to any server that is set up to be accessed through the server they dial. It makes a single server available as a gateway to other Notes servers.

For example, imagine that you are on a business trip to a city in which your company has a remote office. When you get to the hotel in the evening, you want to check your mail. Instead of making the long-distance call to your home office, however, you want to call the local office, as shown in the diagram in Figure 6.27. This situation is an ideal use of Remote Passthru.

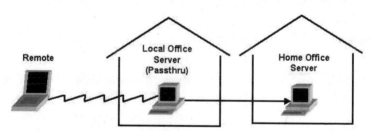

Figure 6.27 Simple Passthru example

To use Passthru in this example, the user would dial up to connect to the server at the local office. That server would act as a stepping stone to the mail server and the database server in the home office. The connection between the local office server and your home office server(s) can be made in a variety of ways, including dial-up and LAN/WAN connection. Either way, the end user will only have to make one call or connection.

To configure Passthru, you must create or edit several documents, including Location, Connection, Server Connection, and Passthru Server documents. Let's look at each type in more detail.

■ *Create a Location document in the user's Personal NAB which specifies the default Passthru server.* The user's NAB has a set of Location documents that specify a variety of information about how the user works in that location. A typical Location document, for example, specifies information including how the user connects to the Notes server (e.g., LAN, dial-up modem, or no connection), how the user accesses and uses the Internet, where the user receives mail, what the user's replication schedule is, and what the user's default Passthru server is. If you used a User Setup Profile when creating this user, some of this information will be completed when the user is configured. If you did not use a User Setup Profile, the default Location documents (Home, Internet, Island, Office, and Travel) exist, but will need to be completed with the additional information necessary for Passthru, as shown in Figure 6.28. If the user always contacts Austin/Server/TotalSem as the Passthru server, for example, this information could be placed in the default Passthru server field for all Locations.

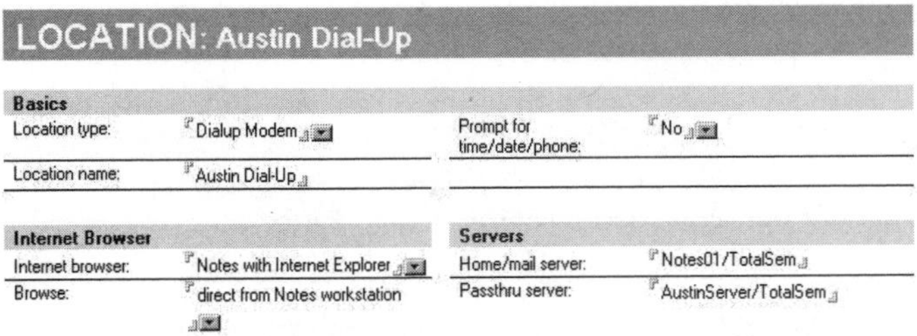

Figure 6.28 Location document specifying default Passthru server

■ *Create a Connection document in the user's Personal NAB, which specifies the path to the Passthru server.* If you used the User Setup Profile to designate a default Passthru server, Notes automatically creates a dial-up modem Connection document specifying the route to the Passthru server. If you did not use a User Setup Profile, however, you must manually create a Connection document in the user's Personal NAB that designates a route to the Passthru server. This action is not necessary if the user has a direct LAN connection to their Passthru server, however. An example of a user's Connection document specifying the route to the Passthru server is shown in Figure 6.29.

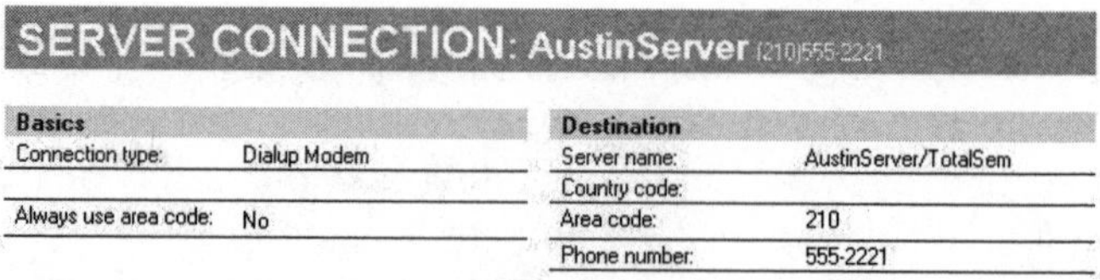

SERVER CONNECTION: AustinServer (210)555-2221

Basics		Destination	
Connection type:	Dialup Modem	Server name:	AustinServer/TotalSem
		Country code:	
Always use area code:	No	Area code:	210
		Phone number:	555-2221

Figure 6.29 Connection document specifying a route to the Passthru server

- *Create a Connection document in the user's Personal NAB, specifying any other Passthru server other than the default.* If the user will be using a Passthru server that is different than the Passthru server designated in their Location document, an additional Connection document is necessary in their Personal NAB—which specifies the Passthru server to use. An example of this type of Connection document is shown in Figure 6.30.

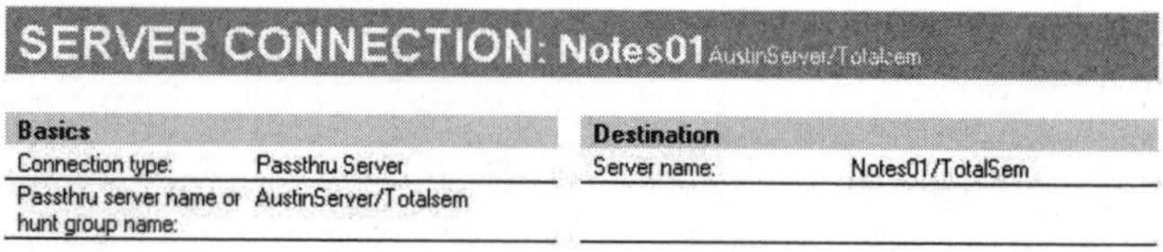

SERVER CONNECTION: Notes01 AustinServer/TotalSem

Basics		Destination	
Connection type:	Passthru Server	Server name:	Notes01/TotalSem
Passthru server name or hunt group name:	AustinServer/Totalsem		

Figure 6.30 Connection document specifying a Passthru server

- *Create additional Location documents in the user's Personal NAB, which specify alternative default Passthru servers.* As an alternative to creating additional Connection documents for different Passthru servers, you can create additional Location documents that specify different Passthru servers.

- *Create Server Connection documents for the Passthru server in the NAB, specifying the route from the Passthru server to the destination servers.* The Passthru server must have a route to the destination server. This route should be specified in a Server Connection document, as shown in Figure 6.31.

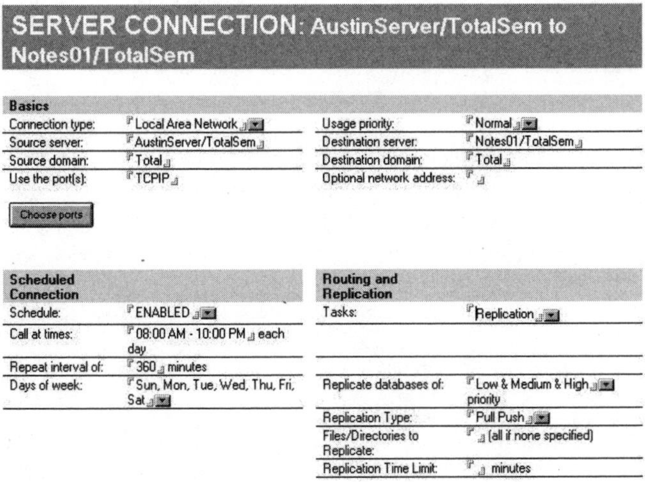

SERVER CONNECTION: AustinServer/TotalSem to Notes01/TotalSem

Basics

Connection type:	Local Area Network	Usage priority:	Normal
Source server:	AustinServer/TotalSem	Destination server:	Notes01/TotalSem
Source domain:	Total	Destination domain:	Total
Use the port(s):	TCPIP	Optional network address:	

[Choose ports]

Scheduled Connection		**Routing and Replication**	
Schedule:	ENABLED	Tasks:	Replication
Call at times:	08:00 AM - 10:00 PM each day		
Repeat interval of:	360 minutes		
Days of week:	Sun, Mon, Tue, Wed, Thu, Fri, Sat	Replicate databases of:	Low & Medium & High priority
		Replication Type:	Pull Push
		Files/Directories to Replicate:	(all if none specified)
		Replication Time Limit:	minutes

Figure 6.31 Server Connection document

- *Edit the Server document on the Passthru server to allow Passthru connections.* Each server that will participate in Passthru must enable Passthru using the fields in the Restrictions section of the Server document, as shown in Figure 6.32.

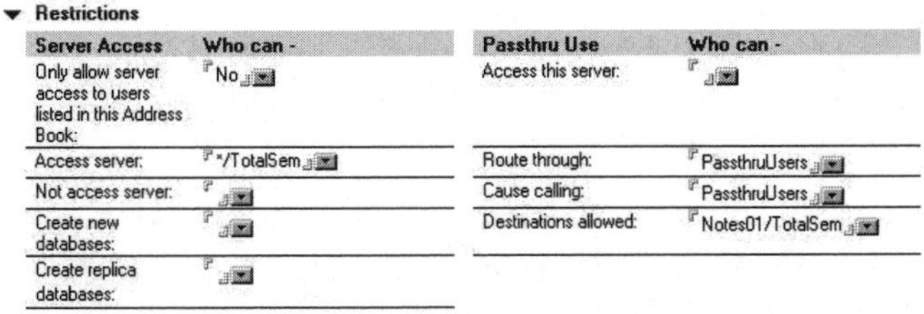

▼ **Restrictions**

Server Access	**Who can -**	**Passthru Use**	**Who can -**
Only allow server access to users listed in this Address Book:	No	Access this server:	
Access server:	*/TotalSem	Route through:	PassthruUsers
Not access server:		Cause calling:	PassthruUsers
Create new databases:		Destinations allowed:	Notes01/TotalSem
Create replica databases:			

Figure 6.32 Passthru Server document

The Passthru server itself must list the users who can force it to call the destination servers in the Route through and Cause calling fields. If a user is not listed in these fields, they cannot use this server for Passthru. The Destinations allowed field must also contain the names of all destination servers that this Passthru server can be used to contact. If you leave this field blank, all destinations are enabled.

> **NOTE**
> It is easiest to use a single group name in these fields, such as PassThruAccess, or RouteThruAccess, to make administration and changes easier on the administrator.

■ *Edit the Server documents for any destination servers to enable them to be accessed via Passthru.* Figure 6.33 shows a Server document for a server that is being accessed using Passthru. In this case, the Access this server field is completed to show the users and servers that can access this server through Passthru. If this field is blank, no user(s) or server(s) can access this server using Passthru.

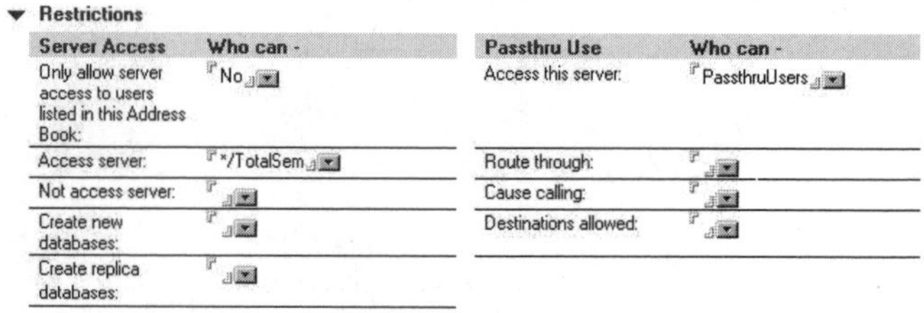

Server Access	Who can -	Passthru Use	Who can -
Only allow server access to users listed in this Address Book:	No	Access this server:	PassthruUsers
Access server:	*/TotalSem	Route through:	
Not access server:		Cause calling:	
Create new databases:		Destinations allowed:	
Create replica databases:			

Figure 6.33 Server document for a server being accessed by Passthru

> **NOTE**
> The feature of Passthru is new to Notes R4.X. For this reason, only R4.X servers can be involved in Passthru. Notes V3.X servers can be involved only in Passthru as final destination servers. They cannot act as Passthru servers, nor can they use Passthru to access another server.

To use the Passthru example described earlier, imagine a situation in which a user on the road wanted to connect to his faraway home server via a local office server. You would create the following documents (assuming that the user did not have the remote office's server as the default):

■ Create a Passthru-type Connection document that specifies the remote server (Vancouver_Notes/TotalSem) as the Passthru server and the home server (Houston_Notes/TotalSem) as the destination server, as shown in Figure 6.34.

■ Create a dial-up modem-type Connection document specifying the phone number and other connection information to reach Vancouver_Notes/TotalSem, as shown in Figure 6.35.

SERVER CONNECTION: Houston_Notes Vancouver_Notes/TotalSem

Basics		Destination	
Connection type:	Passthru Server	Server name:	Houston_Notes/TotalSem
Passthru server name or hunt group name:	Vancouver_Notes/TotalSem		

Figure 6.34 Passthru Connection document

SERVER CONNECTION: Vancouver_Notes 555-4333

Basics		Destination	
Connection type:	Dialup Modem	Server name:	Vancouver_Notes/TotalSem
		Country code:	
Always use area code:	No	Area code:	604
		Phone number:	555-4333

Figure 6.35 Dial-up Connection document

- Vancouver_Notes/TotalSem must have a Connection document specifying its route to Houston_Notes/TotalSem, as shown in Figure 6.36.

SERVER CONNECTION: Vancouver_Notes/TotalSem to Houston_Notes/TotalSem

Basics			
Connection type:	Dialup Modem	Usage priority:	Normal
Source server:	Vancouver_Notes/TotalSem	Destination server:	Houston_Notes/TotalSem
Source domain:	Total	Destination domain:	Total
Use the port(s):	COM1	Destination country code:	
Always use area code:	Yes	Destination area code:	713
		Destination phone number:	555-1222
		Login script file name:	
		Login script arguments:	

Figure 6.36 Vancouver_Notes/TotalSem Connection document

- Vancouver_Notes/TotalSem must enable the user to use it for Passthru, by adding the user to the fields in the Restrictions section of the Server document, as shown in Figure 6.37. In addition, Houston_Notes/TotalSem must be an allowed destination.
- Finally, Houston_Notes/TotalSem must enable access by the user and Vancouver_Notes/TotalSem in its Restrictions section, as shown in Figure 6.38.

▼ Restrictions

Server Access	Who can -		Passthru Use	Who can -
Only allow server access to users listed in this Address Book:	No		Access this server:	
Access server:	*/TotalSem		Route through:	PassthruUsers
Not access server:			Cause calling:	PassthruUsers
Create new databases:			Destinations allowed:	Houston_Notes/TotalSem
Create replica databases:				

Figure 6.37   Vancouver_Notes/TotalSem Server document

▼ Restrictions

Server Access	Who can -		Passthru Use	Who can -
Only allow server access to users listed in this Address Book:	No		Access this server:	PassthruUsers, Vancouver_Notes/TotalSem
Access server:	*/TotalSem		Route through:	
Not access server:			Cause calling:	
Create new databases:			Destinations allowed:	
Create replica databases:				

Figure 6.38   Houston_Notes/TotalSem Server document

EXAM TIP

Remote Passthru is new to Notes R4.X and is therefore tested thoroughly on the System Administration I exam. You should know how to configure Passthru, which documents are required, and which types of servers can use Passthru.

An additional use of Passthru is in a scenario in which servers and users have different protocols. If a user who uses TCP/IP needs to access a server that uses SPX, he cannot do so directly. If, however, another server in the organization that uses both TCP/IP and SPX is configured as his default Passthru server, he can reach his destination server easily. One of the benefits of using Passthru in this situation is that the database that the user is trying to access does not need to be on the Passthru server. Because the Passthru server is acting only as a gateway, or steppingstone, the database needs only to reside on the ultimate destination, as shown in Figure 6.39.

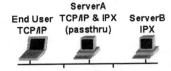

End User   ServerA   ServerB
TCP/IP   TCP/IP & IPX   IPX
    (passthru)

Figure 6.39   Using Passthru to overcome protocol differences

 NOTE
Keep in mind the following limits of Passthru:

- The hard-coded limit for hops is 10, which means that a user cannot go through more than 10 Passthru servers on the way to the final destination. A more practical limit is actually two or three hops.
- As mentioned earlier, a Notes V3.X server cannot be a Passthru server and cannot use a Passthru server. The only way that a Notes V3.X server can be involved in Passthru is as a final destination.

TRACING NETWORK CONNECTIONS

To help an administrator troubleshoot user connections, Notes provides the capacity to trace a network connection. This feature can be helpful if a user is unable to contact a destination. If a user is suddenly unable to contact a destination that could be contacted previously, for example, tracing the connection might tell an administrator if a Connection document was deleted or if a server went down. To trace a connection, select File . . . Tools . . . User Preferences from the menus. Select the Ports panel, as shown in Figure 6.40, from the User Preferences dialog box.

The Ports panel determines and displays which Notes communications ports are enabled and what their options are. In addition, it gives users and administrators the capacity to add, delete, and rename Ports. The Encrypt Network data check box enables the user to encrypt all information that is sent through a particular port.

To trace a connection using a particular port that is already enabled, highlight that port from the list and click the Trace Connection button. The Trace Connections dialog box is displayed, as shown in Figure 6.41. Type the name of the server that you are trying to contact in the Destination or select a Connection that you are trying to trace by clicking the drop-down arrow. The names available in this list are the Connections that exist in the user's Personal NAB.

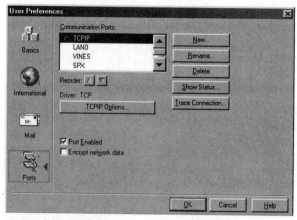

Figure 6.40 Ports information in the User Preferences dialog box

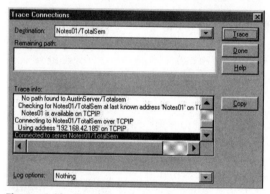

Figure 6.41 Trace Connections dialog box

You can choose from five logging options in this dialog box:

- *Nothing.* Nothing will be shown in the log—including errors—if you choose this option.

- *Errors only.* This setting will show only errors that occur in the trace after the Trace button is clicked. No other information about the path will be reported to the user.

- *Summary Progress Information.* This option shows the Error information that was displayed in Errors Only. This option also reports any major events or comments while connecting to the designated Notes server.

- *Detailed Progress Information.* This setting is the default and returns all of the information available from the Summary Progress Information, as well as additional information about how the connections are made to the designated server, as shown in Figure 6.42.

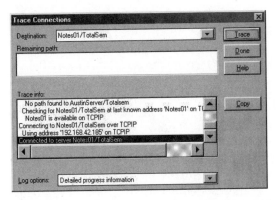

Figure 6.42   Detailed Progress Information for reaching a server

- *Full Trace Information.* This setting returns all of the information available from the Detailed Progress Information, as well as additional comments about the entire process of making the connections to the designated server.

Tracing a connection helps users and administrators understand why a particular connection is or is not working. In addition, it can help a user or administrator understand what method the workstation or server is using to make the connection (i.e., a Connection document, a wild-card connection, or a broadcast or probe). The capacity to trace a connection is not limited by the Notes Domain, the organization, or the NNN.

Maintaining Notes Databases

Occasionally, databases can become corrupted. Documents, views, and folders can all become corrupted in the following types of situations:

- During a server crash or other improper Notes shutdown
- When one or more users open Notes databases through the operating system, instead of through the Notes server

- Where there is improper data access in a Notes application by an outside application

Three basic elements help maintain database integrity and prevent data loss:

- *Run the Notes server program FIXUP to repair damaged databases.* To run FIXUP from the Notes server on any closed databases, type the following command at the server console: LOAD FIXUP Databasename. This program will attempt to repair damaged documents, folders, and other notes in the specified database. On most server platforms, you can also run FIXUP from the operating system, which enables FIXUP to run on all databases—including the ones that are open. For servers running on Windows NT, for example, you would open a command prompt and type the following command: NFIXUP path\databasename arguments. When running FIXUP from a command prompt, specify the entire path to the database —unless the path is already defined by the operating system's PATH= statement. In addition, you can use the following arguments with the FIXUP task:

  - -L reports every database FIXUP and checks for corruption to the Notes Log. This argument is useful only when running FIXUP on all databases.

  - -V prevents FIXUP from checking views. This argument reduces the time required to run FIXUP.

  - -I limits FIXUP to checking only those documents modified since the last time FIXUP ran.

  - -N prevents FIXUP from purging corrupted documents. The next time FIXUP runs or the next time a user opens the database, FIXUP checks the database again. This argument enables you to copy and paste any corrupted documents into another database, as an attempt to salvage the data if no backup or replica of the database exists.

- *Run the Notes server program UPDALL.* The UPDALL server task is listed in the NOTES.INI setting ServerTasksAt2 to run automatically. UPDALL is used to update all views that have been accessed at least once and all full-text indexes for all databases on a server— as well as to rebuild corrupted views and view indexes. If you notice an inability to open a view, data in a document that does not match what is being shown in the view, or unreadable characters in a column in a view, you probably have a corrupt view.

If you use the optional -R parameter with UPDALL, you can use it to find and rebuild corrupted views. For more information on the parameters available with UPDALL, use the Notes Administration Help database. You can also rebuild corrupt views from the workspace by selecting the view in the database and pressing SHIFT+F9.

- Keep a current backup of all Notes databases and configuration files, including all ID files and the NOTES.INI files for the servers. Without backups of your data and configuration files, a server crash can be catastrophic. In the current version, Notes does not offer a backup and restore utility; therefore, you must buy a third-party tool that is specifically designed for use with Notes databases. The backup utility should be able to backup databases and files that are open, or some of the files on your server will never be backed up (such as NAMES.NSF, LOG.NSF, and the server ID file). A common misconception is that you can use replication as a backup solution. Unfortunately, if a user deletes a document that should not have been deleted, that document will be deleted from the replica as well when replication occurs. Additionally, if corruption occurs in your database, that corruption will probably be passed on to the replicas. Make a backup of critical data every day, especially the Public NAB, the Shared mail database (if you are using Shared mail), and the server and certifier ID files. If you have a backup and need to restore a document or other element, the best way to accomplish this is to restore the backup to a separate location with a different name. Copy the documents and other elements into the original database and then delete the copy.

Note that these suggestions and options are not completely fool-proof. If you verify and back up your data regularly, however, you should be able to maintain good data integrity.

Questions

1. Where should Kathy go to begin renaming a user?

    a. The Administration Requests database

    b. The Public NAB

    c. The Server Administration panel

    d. The Certification Log

2. What databases must exist to use the AdminP?

    a. ADMIN4.NSF

    b. CERTBOOK.NSF

    c. CERTLOG.NSF

    d. NAMELOG.NSF

3. Janet is the administrator of a Notes organization that uses both Notes V3 and Notes R4 servers. What must she do first to utilize AdminP in her organization?

    a. Configure AdminP settings and create the required databases. Run AdminP on her servers.

    b. Set the Administration Server of the Public NAB.

    c. Upgrade all her Notes V3 servers to R4.

    d. Create a server Program document to schedule AdminP.

4. Bobby is using AdminP to rename users. He has set up AdminP correctly, and it has been functioning perfectly. The user he is renaming today, however, is impatient. How can he speed up the process?

5. Name two ways to find out the Administration Server for a database.

6. Paul is a new Notes administrator. He wants to monitor the users on his server to understand how they are using the server. What command can he type at the console to find out this information?

   a. SH TA

   b. TELL SERVER SHOW TASKS

   c. SH US

   d. TELL SERVER SHOW USERS

7. Which of the following commands would close the user sessions on Server1 so that Rick can shut it down?

   a. TELL SERVER DROP ALL

   b. TELL SERVER CLOSE CONN

   c. DROP USERS

   d. DROP ALL

8. Judy is listed as the Manager of the Public NAB and is in a group called Administrators. When she tries to type a command at the remote console, however, she gets an error. What might be the problem?

   a. Neither Judy nor the Administrators group is in the Remote-Access field of the Server document.

   b. The Administrators group is misnamed and should be called Admin.

   c. Neither Judy nor the Administrators group has the NetModifier role in the NAB.

   d. Neither Judy nor the Administrators group is in the Administrators field of the Server document.

9. John wants not only to see the results of the commands he types at the remote console, but also to see the other activities occurring on the server. What option should he select?

   a. Pause output

   b. Echo on

   c. Live console

   d. Monitor on

10. CERTBOOK.NSF can be no larger than 1GB. K.C. wants to make sure she knows that the database is approaching its limit when it reaches 750MB. She also wants to make sure that users have to stop putting documents in when the database reaches 800MB. What settings should she put on the database?

    a. Set the database Quota to 800,000KB and the Threshold to 750,000KB.

    b. Set the database Threshold to 750,000KB and the Quota to 800,000KB.

    c. Set the database limit to 750,000KB and the Quota to 800,000KB.

    d. Set the database Quota to 750,000KB and the limit to 800,000KB.

11. Allie wants to make her databases smaller and more efficient by removing any unused white space. What command does she use?

    a. COLLECT

    b. COMPACT

    c. UPDALL

    d. FIXUP

12. Mel wants to try to repair a corrupt database. What command might she use?

    a. COLLECT

    b. COMPACT

    c. UPDALL

    d. FIXUP

13. Sean wants to update the views and view indexes in his databases. What command should he use?

    a. COLLECT

    b. COMPACT

    c. UPDALL

    d. FIXUP

14. Rob uses a laptop with a dial-up modem to connect to his mail server, Server1, when he is at home. Tonight, he realizes that he also needs to replicate the inventory database from Server2. Unfortunately, Server2 does not have a modem. Server1 and Server2 are connected using TCP/IP over the LAN. What should Rob do?

15. Where is James' Passthru server defined?

   a. Location document in the Personal NAB

   b. Location document in the Public NAB

   c. Connection document in the Personal NAB

   d. Connection document in the Public NAB

16. If Steve is using Server1 as his Passthru server to reach Server2, which of the following is correct for the fields on Server1's Server document?

   a. Passthru Use = blank; Route Thru = Server2; Cause Calling = Server2; Destinations Allowed = Server1

   b. Passthru Use = Steve; Route Thru = Steve; Cause Calling = Steve; Destinations Allowed = Server1

   c. Passthru Use = Steve; Route Thru = Server1; Cause Calling = Server1; Destinations Allowed = Server2

   d. Passthru Use = blank; Route Thru = Steve; Cause Calling = Steve; Destinations Allowed = Server2

17. What is the default logging setting in the Trace Connections dialog box?

   a. Nothing

   b. Errors Only

   c. Summary Progress Information

   d. Detailed Progress Information

18. Where do you access the Trace Connections dialog box to trace a route to a server?

    a. File . . . Tools . . . Server Administration . . . Servers

    b. File . . . Tools . . . User Preferences . . . Ports

    c. Public NAB . . . Actions . . . Trace Connection

    d. File . . . Tools . . . Server Administration . . . Administration . . . Trace Connection

19. If you are selecting files to back up, which of these would you select?

    a. NAMES.NSF

    b. PERNAMES.NTF

    c. LOG.NSF

    d. SERVER.ID

20. David has configured a default Passthru server in his Home Location document to send and receive mail on his laptop. He has a Connection document configured to take him to that default Passthru server. He goes home and tries to connect to send and receive mail. He receives an error that says Notes cannot find his mail server. What is wrong?

    a. He did not change the location to Home from Office.

    b. He did not put the correct mail server in his Location document.

    c. He did not create a Connection document specifying the Passthru server.

    d. He did not connect his laptop to a phone line to enable it to dial out.

Answers

1. b. The NAB.

2. a and c. ADMIN4.NSF and CERTLOG.NSF

3. c. Upgrade all her Notes V3 servers to R4.

4. He can force AdminP to begin processing requests by typing TELL ADMINP PROCESS ALL at the console.

5. TELL ADMINP SHOW DATABASES will show you which databases use the current server for the Administration server. Looking at the ACL Advanced panel will show which server a particular database uses for its Administration Server.

6. c. SH US

7. d. DROP ALL. To drop a single user, type DROP and the hierarchical name of the user inside quotes.

8. d. Neither Judy nor the Administrators group is in the Administrators field of the Server document.

9. c. Live console

10. b. Set the database Threshold to 750,000KB and the Quota to 800,000KB.

11. b. COMPACT

12. d. FIXUP

13. c. UPDALL

14. Rob should configure a Passthru Connection document to use Server1 to Passthru to Server2. He should already have a Connection document to specify how to connect to Server1. Passthru will work only if Server1 and Server 2 are configured to enable it.

15. Either a or c. If it is the default Passthru server, it will be defined in a Location document in the Personal NAB. If it is a non-default Passthru server, it will be defined in a Connection document in the Personal NAB.

16. d. Passthru Use = blank; Route Thru = Steve; Cause Calling = Steve; Destinations Allowed = Server2

17. c. Summary Progress Information

18. b. File . . . Tools . . . User Preferences . . . Ports

19. a, c, and d

20. a, b, and d might all be correct.

CHAPTER 7

Advanced Configuration
and Setup

In the first two chapters, we described the process of configuring Notes, creating documents in the NAB, and creating a domain and an organization. We used the domain to route mail, and we used the organization for security. In this chapter, we describe what happens when you work outside your domain or organization. With multiple organizations, you create the need for cross-certification or configuring flat certification. The following is a list of some issues you may encounter with multiple domains:

- Mail routing among domains
- Using multiple address books
- Splitting or merging domains

Before we can discuss any of these issues, however, it may be useful to review both the methods of grouping servers into domains and organizations and the reasons for choosing these groupings.

Objectives

After reading this chapter, you should be able to answer questions based on the following objectives:

- Allow access to your organization through cross-certification.
- Gain access to a flat organization through nonhierarchical certificates.
- Configure routing and replication among domains.
- Monitor routing and replication among domains.
- Create and maintain Cascading Address Books.

Grouping Servers and Users

We usually group Notes servers and users into domains and organizations, based on both the structure and business needs of companies. The following section provides a review of the topics of Notes domains and organizations, discussed in Chapter 1, "Installation and Configuration."

A *domain* is a grouping of Notes users and servers that share a common NAB. The following points describe the main issues that you need to remember about Notes domains:

- The main function of a Notes domain is mail routing.
- All Notes servers and users belong to a domain. The NAB determines their domain.
- Mail routing within a domain is a simple process requiring only a common NNN or a Connection document.
- The domain name usually will represent the company's name and identity to the outside world.
- Using Total Seminars as an example company, the domain could be @TotalSeminars. Total Seminars would have a NAB called Total-Seminars' Address Book.

A Notes *organization* is a grouping of Notes users and servers that are created using a common certifier. This certifier gives those users and servers a common certificate. This in turn gives them the authorization and the capacity to authenticate with one another. The following points describe the main issues that you need to remember about Notes organizations:

- The main function of a Notes organization is security.
- An organization can be either flat or hierarchical in Notes.

- Hierarchical organizations can use both a top-level (organization) certifier and mid-level (organizational unit) certifiers to create users and servers.

- Organizations and organizational units are called certificates. All the certificates for an organization are stored in the NAB.

- These certificates help determine the name of a user or server within an organization.

- The organization name will usually represent how the company is known internally.

- Using Total Seminars as an example company, the organization name could be /TotalSem.

- You combine the certificates with the users' or servers' names to create a distinguished name. Separate the elements of a distinguished user name with slashes. A user in this organization could be called Libby Schwarz/TotalSem or Libby Schwarz/Instructors/TotalSem. A server in this organization could be called ServerA/TotalSem or ServerA/HOU/TotalSem.

- Distinguished names can either be canonical or abbreviated in format. A canonical format includes the name of the component in the distinguished name (for example, CN=Libby Schwarz/OU=Instructors/O= TotalSem). Names are stored internally using the canonical format. Names are displayed using the abbreviated format, which removes the component name (for example, Libby Schwarz/Instructors/TotalSem).

EXAM TIP

Although this review might seem to duplicate the information in Chapter 1, "Installation and Configuration," the System Administration II exam focuses on the concepts of Notes domains and organizations and the differences between the two.

Now that we have reviewed the concepts of Notes domains and organizations, we can discuss some of the options that exist in the relationship between domains and organizations. For most companies, it makes the most sense for one domain to contain one organization. A domain may contain multiple organizations, however, and an organization may contain multiple domains.

When one domain contains one organization, the company name will probably be used for both the domain and organization. For example, the company Total Seminars may use the domain TotalSem and the organization certifier TotalSem. A distinguished name in this environment would look like the example in Figure 7.1.

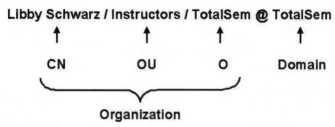

Figure 7.1   Components of a distinguished name

Mail addressing would use both the hierarchical name and the single domain (for example, Libby Schwarz/Instructors/TotalSem @ TotalSem).

One domain can also contain multiple organizations, and this can be set up for a variety of reasons:

- Using multiple organizations to control server access
- Using multiple organizations to enable each distinct business unit within a company to have its own internal identity, but presenting a uniform external identity

The chart in Figure 7.2 shows an example of the company Total Seminars, which uses the domain TotalSem but also uses the two organizations of Authors and Instructors.

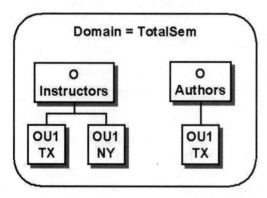

Figure 7.2   One domain, multiple organizations

The fully distinguished names of two example users in these organizations would look like the example in Figure 7.3.

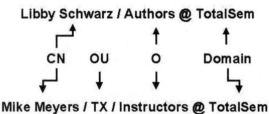

Figure 7.3 Distinguished names using multiple organizations

Mail addressing in such a company would continue to be within a single domain. If Libby Schwarz wanted to send mail to Michael Meyers, for example, the mail addresses would look like these:

```
Libby Schwarz / Authors @ TotalSem
Michael Meyers / TX / Instructors @ TotalSem
```

Finally, an organization can contain multiple domains. If a company has business units that function separately in the external world, the business units might want to be known by different domain names but remain part of the same organization. This setup can be done for a variety of reasons, including:

- The address book has became too large and difficult to maintain.
- The address book has became so large that it is difficult to open and use.
- Administration of the address book needs to be distributed among the business units.

Now, all the users and servers in the organization can authenticate with each other, based on the common top-level certifier. Mail routing will still be defined by Connection documents.

The chart in Figure 7.4 shows an example of the company Total Seminars, which uses the domains Authors and Instructors but uses the organization TotalSem.

The fully distinguished names of the two example users in this organization would look like the example in Figure 7.5.

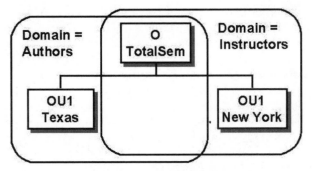

Figure 7.4   One organization, multiple domains

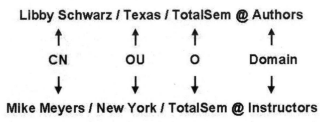

Figure 7.5   Distinguished names using multiple domains

Mail addressing in this style of organization is now outside a single domain. If Libby Schwarz wanted to send mail to Michael Meyers, for example, the mail addresses would look like these:

```
Libby Schwarz / Texas / TotalSem @ Authors
Michael Meyers / New York / TotalSem @ Instructors
```

A Connection document specifying a route between the domains would have to exist in the NABs of each domain for her mail to get to Michael.

Now that we have described the possible scenarios available with multiple organizations and domains, the next sections of this chapter will describe how to allow users who are in different organizations and domains to communicate and share data.

Multiple Organizations

Sometimes users in different organizations need to share information and databases. When they share databases that exist in different organizations, those users will need to be able to authenticate with the other user's organization. Authentication, as you recall, is the capacity to establish a user's or server's identity using the certificates contained in the ID files. When users need to authenticate outside their organization, therefore, the organizations must share certificates. The process of sharing certificates to enable authentication among users and servers that were not created with a common certificate is called *cross-certification*. Additionally, when users communicate outside their organization, additional security concerns need to be addressed—such as connections to the other organization, firewalls, and server security. Finally, we should mention that sometimes other organizations do not use hierarchical naming. In this case, you need to configure flat certification to communicate with those organizations. This section describes cross-certification, how it is configured, the additional security concerns when connecting outside your organization, and connecting to flat organizations.

Cross-Certification

Cross-certification requires each organization to obtain a certificate from the other organization, because they do not already share a hierarchical certificate to enable them to authenticate. When we described distinguished names earlier, we described a distinguished name as containing the certificates held by a user or server. We also described the registration process for users and servers, where the administrator chooses the hierarchical certificates from the appropriate certifier that should be included or inherited in the user's or server's ID file. When users and servers are in different organizations, they are not registered using a common certifier ID. You can say that they have no ancestors, or certificates, in common. Authentication depends on having a certificate in common. Therefore, by default, users and servers in the same organization can authenticate. Users and servers in different organizations must be given a common certificate before they can authenticate.

Cross-certification begins when one company or organization, Company A, obtains an ID file from the other company, Company B. Company A then cross-certifies the ID file. The certificate that is created by the process of cross-certification is stored in Company A's NAB. Company A then sends an ID to Company B. Company B performs the process of cross-certification. The certificate created by Company B is stored in Company B's NAB. Both companies must have a certificate that relates to the other company in their NAB before authentication can occur.

NOTE

Cross-certification is a two-way process that requires both organizations to cross-certify and store a certificate. If the certificate does not exist in either company's NAB, authentication cannot occur.

The process of cross-certification creates a cross-certificate document in the NAB. An example of a cross-certificate document is shown in Figure 7.6.

```
┌─────────────────────────────────────────────────────┐
│  Edit Certificate                                     │
├─────────────────────────────────────────────────────┤
│  CROSS CERTIFICATE:/Instructors/TotalSem              │
│  -/ACMECorp                                           │
│                                                       │
│  Basics                                               │
│  Certificate type:    Notes Cross-Certificate         │
│  Issued By:           /Instructors/TotalSem           │
│  Issued To:           /ACMECorp                       │
│  Combined Name:       OU=Instructors/O=TotalSem:O=A   │
│                       CMECorp                         │
│  Comment:                                             │
│  Organizations:       O=TotalSem:O=ACMECorp           │
└─────────────────────────────────────────────────────┘
```

Figure 7.6 Cross-certificate document

When two organizations have cross-certified, the following results occur:

- Each organization completes the process of cross-certification on an ID that it obtains from the other organization.

- Each organization stores the cross certificate that it creates in the NAB.

- No user ID or server ID is altered.

- Individual users store cross certificates in their Personal Address Book. Servers store cross certificates in the Public Address Book.

Sometimes administrators also need to know what will not occur with cross-certification, including the following items:

- Cross-certification will not work with flat certificates or organizations. Both organizations must use hierarchical naming to cross-certify.

- Cross-certification does not give the other organization the capacity to pass on your certificate and give other organizations the capacity to authenticate with you.

- Cross-certification does not alter your hierarchical name, your structure, or any user or server ID files in your organization.

- Cross-certification does not necessarily give access to all of your servers and databases to the cross-certified organization.

 NOTE
If you communicate with a flat organization, you must obtain flat certificates from it, because you cannot cross-certify. Communicating with flat organizations is discussed later in the chapter.

The process of cross-certification can occur between different levels of the two organizations. The reason that you choose to cross-certify at these different levels is related to the security that you wish to have between the organizations. If you cross-certify from organization ID to organization ID, for example, you open up your entire organization and all your servers to the other organization. If you cross-certify from server ID to user ID, however, you make a limited cross-certification that enables only limited access, which is more secure.

Cross-certification can occur at the following levels:

- *Certifier to certifier.* Organization ID to organization ID, organization ID to organizational unit ID, or organizational unit ID to organizational unit ID.

- *Certifier to server or user.* Organization ID or organizational unit ID to server ID, or organization ID or organizational unit ID to user ID.

- *Server/user to server/user.* Server ID to server ID or server ID to user ID.

CERTIFIER TO CERTIFIER CROSS-CERTIFICATION

When you cross-certify between two certifiers, you are using open security. If you cross-certify at the top of your organization, from the organization ID of Company A to the organization ID of Company B, any user or server from either organization can authenticate with any user or server from the other organization, as shown in Figure 7.7.

If you cross-certify from the organization ID of Company A to the organizational unit ID of Houston/Company B, anyone in Houston/Company B can authenticate with any user or server certified by the Company A organization ID. Users or servers certified by Company A can authenticate with users or servers certified by Houston/Company B, as shown in Figure 7.8.

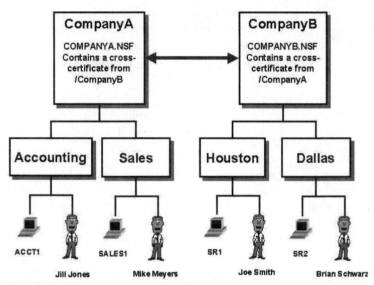

Figure 7.7   Cross-certifying at the O level

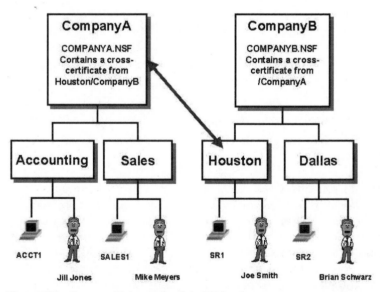

Figure 7.8   Cross-certifying at the O to OU level

If you cross-certify from the organizational unit ID of Accounting/Company A to the organizational unit ID of Houston/Company B, the users and servers certified by Accounting/Company A can authenticate with the users and servers certified by Houston/Company B, as shown in Figure 7.9.

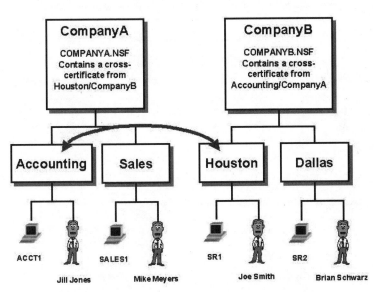

Figure 7.9 Cross-certifying from OU to OU

CERTIFIER TO SERVER OR USER CROSS-CERTIFICATION

If you cross-certify the organization ID of Company A to the server ID of Srv1/Houston/Company B, Srv1/Houston/Company B can authenticate with all the users and servers in Company A. The users and servers in Company A, however, can authenticate only with Srv1/Houston/Company B. This example is shown in Figure 7.10.

If you cross-certify the organizational unit ID of Accounting/Company A to the server ID of Srv1/Houston/Company B, Srv1/Houston/Company B can authenticate with the users and servers certified by Accounting/Company A. The users and servers certified by Accounting/Company A can authenticate with Srv1/Houston/Company B, as shown in Figure 7.11.

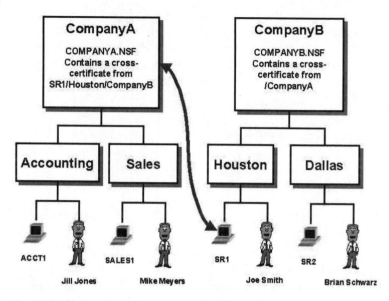

Figure 7.10   Cross-certifying from O to Server

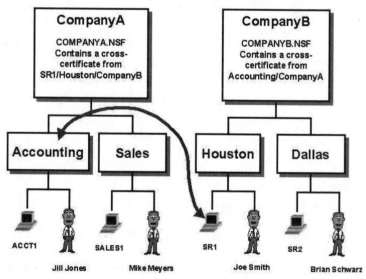

Figure 7.11   Cross-certifying from OU to Server

If you cross-certify the organization ID of Company A to the user ID of Joe Smith/Houston/Company B, Joe Smith/Houston/Company B can authenticate with all of the users and servers certified by Company A. No users from Company A can authenticate with Company B. This option is displayed in Figure 7.12.

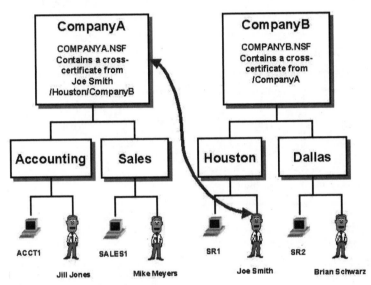

Figure 7.12   Cross-certifying from O to User

If you cross-certify the organizational unit ID of Accounting/Company A to the user ID of Joe Smith/Houston/Company B, Joe Smith/Houston/Company B can authenticate with all of the users and servers certified by Accounting/Company A. No users from Company A can authenticate with Company B. This option is displayed in Figure 7.13.

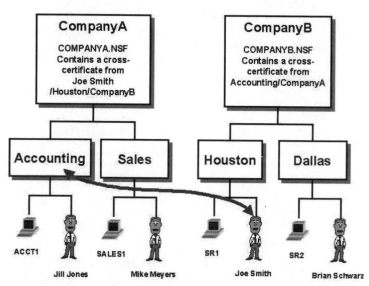

Figure 7.13   Cross-certifying from OU to User

SERVER TO SERVER AND SERVER TO USER CROSS-CERTIFICATION

If you cross-certify the server ID of Acct1/Accounting/Company A to the server ID of Srv1/Houston/Company B, these two servers can authenticate with each other. No other users or servers in Company A can authenticate with any other user or server in Company B, as shown in Figure 7.14.

If you cross-certify the server ID of Acct1/Accounting/Company A to the user ID of Joe Smith/Houston/Company B, Joe Smith/Houston/Company B can authenticate to Acct1/Accounting/Company A, as shown in Figure 7.15. No other users from Company B can authenticate to Acct1/Accounting/Company A. No users from Company A can authenticate to Company B.

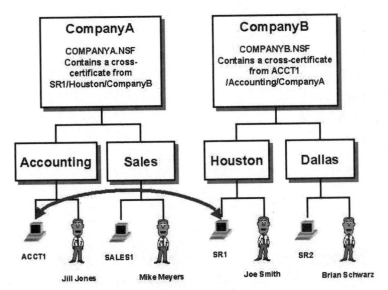

Figure 7.14   Cross-certifying from Server to Server

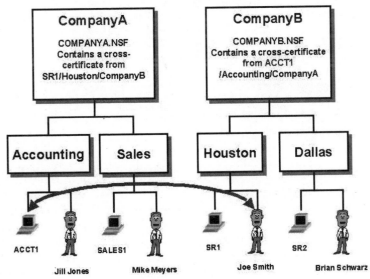

Figure 7.15   Cross-certifying from Server to User

HOW TO CROSS-CERTIFY WITH ANOTHER ORGANIZATION

Lotus Notes offers four methods of cross-certifying between organizations:

- Using Notes Mail
- Using a disk
- On demand
- Verbally

To use Notes Mail to request a cross certificate, select File . . . Tools . . . User ID from the menus. Select the Certificates panel as shown in Figure 7.16.

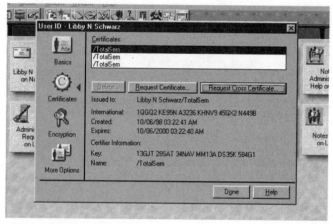

Figure 7.16   User Certificates panel

When you click the Request Cross Certificate button, you will be prompted to choose the ID that should be cross-certified. You can choose your user ID, an organization ID, an organizational unit ID, or a server ID from the Choose ID to Be Cross-Certified dialog box. After you choose the ID to be cross-certified, Notes displays the Mail Cross Certificate Request dialog box, shown in Figure 7.17. In the To: field, type the address of the administrator in the other organization, or select a name using the Address button.

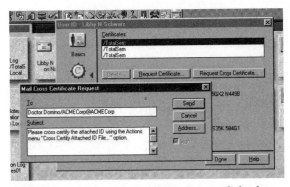

Figure 7.17 Mail Cross Certificate Request dialog box

 NOTE
When using Notes Mail to cross-certify two organizations, some way to send mail between the organizations must already exist.

When you receive this type of request in Notes Mail, open the request and choose Actions . . . Cross Certify Attached ID File from the menus. Notes displays the Choose Certifier ID dialog box, enabling you to select the certifier to use to cross-certify the ID file. This certifier can be a server, organizational unit, or organization certifier. After you decrypt the ID using the correct password, Notes displays the Cross Certify ID dialog box, shown in Figure 7.18.

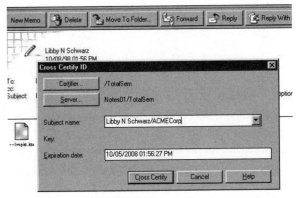

Figure 7-18 Cross Certify ID dialog box

Use this dialog box to confirm the correct certifier, certification server, object to be certified, and certificate expiration date. Then click Cross Certify. A certificate is added to the NAB.

Remember that cross-certification must be two-way for authentication to occur. These steps must be performed in each organization.

You may also request and create cross certificates using a disk. The first step to request a cross certificate from another organization using this method is to create a safe ID, put it on a floppy diskette, and mail it to the administrator in the other organization. A *safe ID* is a copy of an ID file that can be used only for cross-certification functions. A safe ID cannot be used to perform any other Notes functions. A user could create a safe copy of their own ID for cross-certification, or an administrator might create a safe copy of the organization ID, an organizational unit ID, or a server ID. To create a safe copy of an ID file, choose File . . . Tools . . . Server Administration from the menus. When you are in the Server Administration panel, select Administration . . . ID File . . . from the menus, as shown in Figure 7.19.

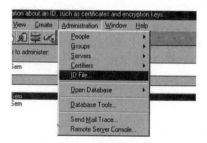

Figure 7.19 Select an ID file.

Use the Choose ID File to Examine dialog box to select the ID file for which you want to request a cross certificate. After you decrypt the ID file by typing in the correct password, the ID file dialog box is displayed. Select the More Options tab and choose Create Safe Copy, as shown in Figure 7.20.

By default, Notes names all safe copies of ID files SAFE.ID. You may want to use a more specific name that includes the name of the original ID. Select a location to store the safe ID. By default, Notes tries to put it on a floppy diskette, which is generally a good idea.

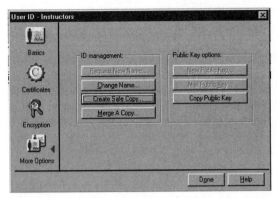

Figure 7.20   Create a safe ID

After creating the safe ID, you will send or give the ID file on a disk to the administrator at the other organization. When you receive a safe copy of an ID file that needs to be cross-certified, choose File . . . Tools . . . Server Administration from the menus. From the Administration menu, choose Certifiers . . . Cross Certify ID file. Notes prompts you to select an ID file to use to certify the safe ID. Then, Notes prompts you to select the ID that should be cross-certified. After you select the appropriate safe ID, Notes displays the Cross Certify ID dialog box.

Again, remember that cross-certification must be two-way for authentication to occur. Administrators in each organization must perform these steps.

Cross-certification on demand occurs when a user in one organization attempts to access a server in another organization, and the user is not cross-certified. You may also use cross-certification on demand when you open mail that has been signed (as discussed in Chapter 3, "Notes Security") by a user in another organization— and you are not cross-certified. You receive the following error message from Notes in this situation:

```
Your local Address Book does not contain a cross-certificate
for the organization named below. You therefore can't be sure
that documents signed by its members are authentic or that you
are actually communicating with its servers unless you can ver-
ify that the key below is correct. Would you like to suppress
this warning in the future by creating a cross-certificate for
this organization?
```

If the warning message appears, either you are not cross-certified with the other organization or you have a cross certificate, but it is not stored in the Public Address Book. The cross certificate may be stored in your Personal Address Book, for example, where other users cannot access it. If this is the case, paste a copy of the cross certificate into the Public Address Book. Similarly, your organization may contain a cross certificate in the Public NAB that you may need to copy to your Personal NAB.

Notes displays the name of the organization and its public key, as well as Yes, No, and Cancel buttons that enable you to do one of the following actions:

- Choosing Yes enables you to cross-certify the O of the user or server. The cross certificate will be placed in your Personal NAB. Note that you can access a server only if that server has cross-certified you, or if your organization or server enables anonymous access.

- Choosing No does not create a cross certificate with the other organization. Notes then offers you unauthenticated access to the server if allowed, or it offers the capacity to read the signed message without verifying the signature. The original error message described earlier appears each time you connect to a server in that organization or read a signed message from a user in that organization.

- Choosing Advanced Options enables you to create a cross certificate for the server or sender of the message, or for any of its certifiers. If you cross-certify a certifier rather than a user or a server, and you place the cross certificate in a Public Address Book, you can suppress the error message for all members of the organization. Note that you can only access a server if that server has cross-certified you, or if your organization or server enables anonymous access.

NOTE

To avoid the possibility of cross-certifying someone incorrectly when using On Demand, call someone from the other organization and ask them to read you their organization's public key. (Users in an organization can verify their organization's public key by examining the Certificates pane of their User ID dialog box.) Compare this public key with the validation code on the Cross Certify ID dialog box.

When you need to cross-certify an ID verbally, you must contact the administrator at the other organization. The administrator should open the Server Administration panel using File . . . Tools . . . Server Administration. From the Server Administration panel, the administrator must select Administration . . . Cross Certify key. A Cross Certify Key dialog box similar to the Cross Certify ID dialog box is displayed, as shown in Figure 7.21.

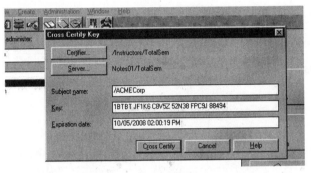

Figure 7.21 Cross Certify Key dialog box

You must then tell the other administrator two pieces of information:

- The name of the ID to be certified, exactly as it appears on the Basics tab of the ID dialog box shown in Figure 7.22.

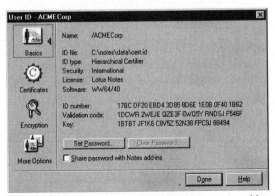

Figure 7.22 ID dialog box showing the name and key necessary for verbal cross-certification

■ The public key of the ID to be certified, exactly as it appears on the Basics tab of the ID dialog box shown in Figure 7.22.

The administrator types this information in the Subject Name and Key fields of the Cross Certify Key dialog box.

Again, remember that cross-certification must be two-way for authentication to occur. Administrators in each organization must perform these steps.

NOTE

All these methods require the person who issues the cross certificates to have Editor access or higher to the NAB and to have the access and password necessary to open the certifier ID file that will be used in the cross-certification process.

Additional Security Issues with External Connectivity

When you begin communicating with servers and users outside your organization, there are some additional security and connectivity issues. Although these issues are not serious considerations for the System Administration II exam, they are significant in terms of understanding how you connect to another organization.

■ *How is the connection established?* Notes servers can use a variety of methods to establish connections to remote servers and organizations. These methods of connecting include direct modem connections (either analog or digital); a private network to which both companies belong, such as those provided by IBM, CompuServe, and MCI; or a connection through a public network, such as the Internet. The main differences between these types of connections are the speed and amount of security they provide.

■ *What protection is being used to secure the connection?* Some connection types, such as direct connection through a modem, enable the use of all levels of Notes security—because only Notes servers and clients can dial directly into a Notes server. When making an indirect connection, such as through a private or public network, additional security measures such as firewalls and encryption should be considered. Firewalls are servers that sit between your Notes servers and the external networks. These servers examine all the data that passes through, using filters and rules to determine whether the data should be allowed to pass. Notes does not provide a specific firewall application. You should use one, however, to protect your Notes environment. In addition, encryption is available in Notes for any port that you use for communication.

■ *Notes server access security.* After successful authentication takes place between two servers or a user and a server, the security and restrictions sections of the Notes Server document provide additional security options that must be passed before the user or server can gain access to the destination server. In addition, the NOTES.INI provides some security settings to limit the access farther. Table 7.1 describes the pertinent security and restrictions settings in the Server document (as shown in Figure 7.23). For more information on the available NOTES.INI settings, refer to the Notes Administration Help database.

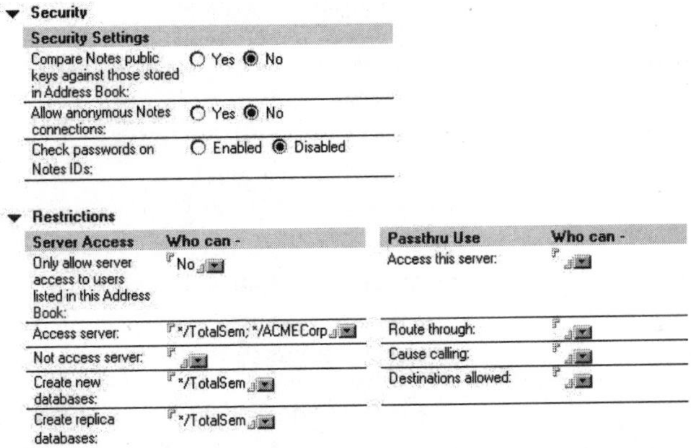

Figure 7.23 Notes Server document Security and Restrictions settings

Table 7.1 Notes Server Document Security and Restrictions Settings

Notes Server Document Security Settings	Description and Use
Compare Notes public keys against those stored in Address Book:	Choose yes to verify public keys against those in the Public Address Book, to help prevent unauthorized access to a server from users who are not listed in the NAB.

continues

Table 7.1 Continued.

Notes Server Document Security Settings	Description and Use
Allow anonymous Notes connections:	Choose yes to enable user access to the server, even if the server cannot authenticate the users. You must also create an entry called Anonymous in database ACLs for databases to which Anonymous users should have access. Assign the appropriate access level— typically Reader access. If you do not add Anonymous as an entry in the ACL, users accessing the server anonymously are given the Default Access.
Only allow server access to users listed in this Address Book:	Choose Yes to prevent users not listed in the Public Address Book from accessing this server. This option prevents users from outside organizations from using this server. If you set this field to Yes, you must specify the group LocalDomainServers or specific server names to give the servers in the NAB access to this server.
Access server:	Specify the users, servers, and groups allowed to access the server. If this field is blank, all certified users and servers can access this server.
Not access server:	Specify the users, servers, and groups that should be denied access to the server.
Create new databases:	Specify the users, servers, and groups allowed to create databases on the server. If this field is blank, all certified users can create new databases.

Notes Server Document Security Settings	Description and Use
Create replica databases:	Specify the users, servers, and groups allowed to create replica databases. If this field is blank, only someone working directly at the server machine can create replica databases.

Connecting to Flat Organizations

Version 2 of Lotus Notes uses flat organizations. As described earlier, flat organizations do not have hierarchical names or certifiers. Instead, users and servers are named using only their common names. Although hierarchical naming is both the recommended and the default naming scheme in Notes, Lotus provides flat naming in V3 and R4 for backward compatibility.

In hierarchical organizations, only the original certificate that stamped or certified an ID is stored in the ID file. All other certificates, such as cross certificates, are stored in either the Public or Personal NAB. In addition, a user, server, or certifier ID can have only one hierarchical certificate, even if this certificate contains information about many levels of ancestors. In flat organizations, however, ID files must contain flat certificates for each server with which they need to authenticate.

If a hierarchical organization needs to communicate with a flat organization, the hierarchical organization must create a flat certificate for its servers. The hierarchical organization must send safe copies of its server IDs with these flat certificates to the flat organization, and vice versa. Both must then present the trusted flat certificates when authenticating with servers in the other organization. Note that each server ID must be treated separately in a flat environment.

Multiple Domains

Earlier, we described the following issues to consider when you must communicate outside your own domain:

- Mail routing among domains
- Using multiple address books
- Splitting or merging domains

To discuss these issues, we must first describe the types of domains that might exist—and then discuss how mail would be routed across these domains. After describing mail routing among the domains, we will also discuss some other issues that arise out of separate domains, such as the need to share Public Address Books and the need to split or merge domains.

Mail Routing Among Domains

The types of domains that are important for the topics covered in the System Administration II exam are adjacent domains and non-adjacent domains.

MAIL ROUTING AMONG ADJACENT DOMAINS

Domains that are adjacent to your current domain have a direct physical connection to your current domain, such as through modems, bridges, routers, and public or private networks. To route mail to an adjacent domain, you must have a Connection document in your NAB that describes a path from some server in your domain to some server in the adjacent domain. Mail can then be transferred between the two domains using the servers that are connected, as shown in Figure 7.24.

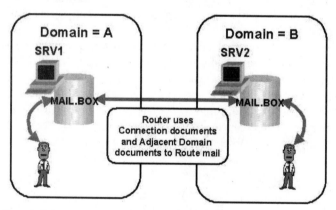

Figure 7.24   Routing mail to an adjacent domain

To define an adjacent domain, create an adjacent domain document in your Public NAB, as shown in Figure 7.25.

DOMAIN: Paris		

Basics		Restrictions	
Domain type:	Adjacent Domain	Allow mail only from domains:	
Adjacent domain name:	Paris	Deny mail from domains:	
Domain description:	The Paris Domain		

Figure 7.25   Adjacent domain document

When user Christian in domain Battlefield needs to send a message to user Roxanne in domain Paris, for example, the router in domain Battlefield examines the address in the mail message, which looks like this: To: Roxanne @ Paris. The router checks for a domain document that defines the domain Paris. When it finds an adjacent domain document for domain Paris, it checks for a Connection document that specifies how and when to transfer mail to a server in domain Paris. Figure 7.26 shows that Christian's mail server, Server A, connects directly to Roxanne's mail server, Server B.

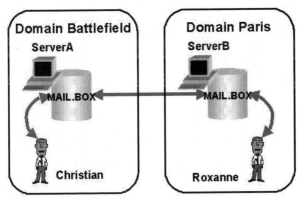

Figure 7.26   Mail routing between adjacent domains using the users' mail servers

When routing mail among domains, however, mail can be routed using any server in either domain that has a Connection document to the other domain. Figure 7.27, for example, shows that Christian's mail server, Server A, does not have any Connection documents that enable it to connect to Roxanne's mail server, Server B. The router finds a Connection document, however, that shows a path from Server C to another mail router in domain Battlefield, to Server D, another mail router in domain Paris.

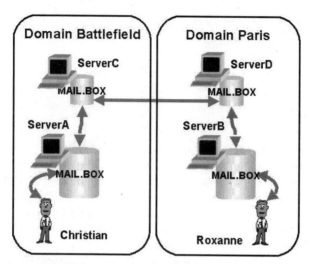

Figure 7.27   Mail routing between adjacent domains using other servers in the domains

When you address messages to be sent to users in adjacent domains, you must use an explicit name to address the message. An explicit name includes not only a distinguished name, but also the domain. For example, you could address a message to Roxanne Smith/Executives/Paris @ Paris.

MAIL ROUTING AMONG NONADJACENT DOMAINS

Domains that are nonadjacent, however, do not have a physical connection. They cannot use modems or networks to connect directly to the other domain. There cannot be a Connection document between the domains for mail routing, because there is no way to connect to that domain. In the case of a nonadjacent domain, an adjacent domain must be identified that can pass the mail messages between the two disconnected domains. Mail is transferred between two nonadjacent domains using a domain that is adjacent to both, as shown in Figure 7.28.

To define a nonadjacent domain, create a nonadjacent domain document in the Public NAB, as shown in Figure 7.29. An adjacent domain and a Connection document will also have to exist for mail to route.

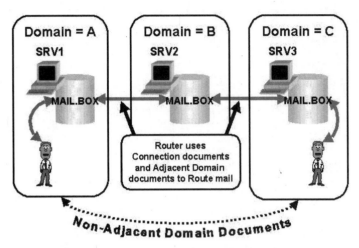

Figure 7.28　Routing mail to a nonadjacent domain

DOMAIN: Paris			
Basics		**Restrictions**	
Domain type:	Non-adjacent Domain	Allow mail only from domains:	
Mail sent to domain:	Paris	Deny mail from domains:	
Route through domain:	Cyrano		
Domain description:	Use Domain Cyrano to send mail to domain Paris.		

Figure 7.29　Nonadjacent domain document

In our previous example, when user Christian in domain Battlefield needs to send a message to user Roxanne in domain Paris, the router in domain Battlefield examines the address in the mail message, which looks like this: To: Roxanne @ Paris. The router checks for a domain document that defines the domain Paris. In this case, it finds a nonadjacent domain document for domain Paris that specifies that any mail sent to domain Paris should be routed through domain Cyrano. The router then searches for an adjacent domain document for domain Cyrano and a Connection document for a server in domain Cyrano. Figure 7.30 shows that Christian's mail server in domain Battlefield, Server A, connects to Server Horse in domain Cyrano. Server Horse in domain Cyrano then connects directly to Roxanne's mail server, Server B, in domain Paris.

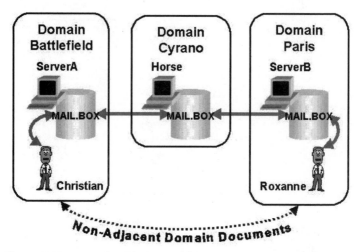

Figure 7.30   Example of routing mail among nonadjacent domains

In the earlier example, the following documents need to exist for mail routing to occur:

- The NAB in domain Battlefield must have a nonadjacent domain document for domain Paris, specifying that mail to Paris routes through Cyrano.

- The NAB in domain Battlefield must have an adjacent domain document for domain Cyrano.

- The NAB in domain Battlefield must have a Connection document that specifies a mail routing connection between Server A and Server Horse.

- The NAB in domain Cyrano must have an adjacent domain document for domain Paris.

- The NAB in domain Cyrano must have a Connection document that specifies a mail routing connection between Server Horse and Server B.

![NOTE icon] **NOTE**

In either case of adjacent or nonadjacent domain routing, for Roxanne to respond to Christian, another set of documents must exist that specifies the route back to domain Battlefield, because mail routing is only one-way. This set of documents includes adjacent and/or nonadjacent domain documents in the NAB in domain Paris, specifying either Battlefield or Cyrano and a mail-routing connection.

When you address a message to a user in an adjacent domain, you use an explicit name such as Roxanne Smith/Executives/Paris @ Paris. When you address a message to a user in a nonadjacent domain, you can use the same address *if* you have a nonadjacent domain document specified in the NAB, as described earlier. If, however, you do not have a nonadjacent domain specified, you must use the explicit name to define both the adjacent and nonadjacent domains to use for routing mail. If you do not have a nonadjacent domain document for domain Paris, for example, you could address a message like the following example: Roxanne Smith/Executives/Paris @ Paris @ Cyrano. The router reads the domain names from right-to-left, in the order that it will transmit them. In this example, the message will first be sent to domain Cyrano, and then Cyrano will send the message to domain Paris.

CONTROLLING MAIL ROUTING AMONG DOMAINS

Administrators in each domain who participate in mail routing among domains have the ability to reject mail coming from a certain domain. You would probably do this activity to avoid overworking resources on your servers. Assume, for example, that you are the administrator of domain Cyrano, and you no longer want to route mail from domain Battlefield to domain Paris. You would place Battlefield in the Deny mail from Domains field of the adjacent domain document that defines the relationship between Cyrano and Paris, as shown in Figure 7.31.

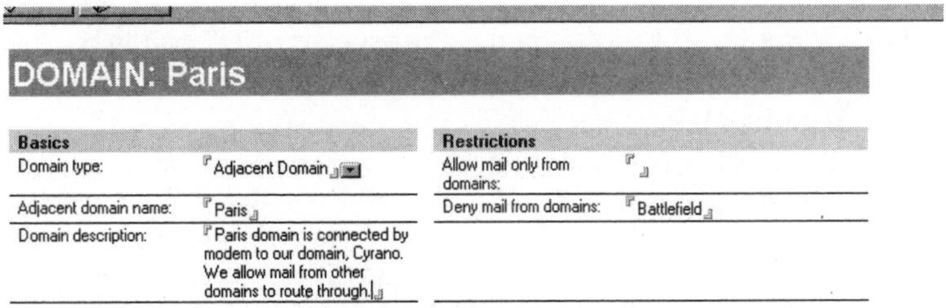

Figure 7.31   Deny mail from domain.

Cascading Public NABs

When you are routing mail among domains in Notes R4, you have two options for the address books in those domains:

- Use one Public NAB per domain, and each server stores only its own domain's NAB.

- Use multiple (cascading) Public NABs, and each server stores the NAB for all domains.

The first option is the default. Each domain contains only its own unique NAB. The NAB is replicated to each server in the domain, and each replica is called NAMES.NSF. A diagram of this option is shown in Figure 7.32.

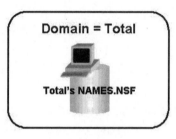

Figure 7.32   One NAB per domain

The second option enables each domain to contain multiple address books. This feature is known as using cascading address books. Each domain will continue to use its own NAB, and this NAB will continue to be replicated to each server in the domain. In addition, each domain will also contain replicas of the Public Address Book of the other domains involved. The domain's own NAB will continue to be called NAMES.NSF, and the other address books will be given unique names that specify their domains. Figure 7.33 shows a diagram of the unique names that would be used with cascading NABs.

When cascading NABs are used, Notes uses all available address books to search for names. The search is done sequentially, as determined by the NAMES= line of the NOTES.INI file. If the line in the NOTES.INI looks like this, for example, NAMES=NAMES, CYRANO, PARIS, the mailer and router will search the address books in the specified order. A diagram of the sequential search is shown in Figure 7.34.

Cascading NABs offer the following major benefit to your users:

Users do not have to use explicit naming when sending mail to users in any of the address books that are cascaded.

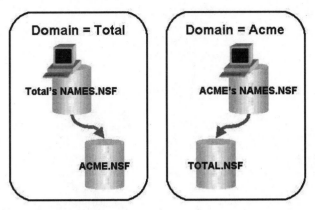

Figure 7.33   NAB filenames with cascading

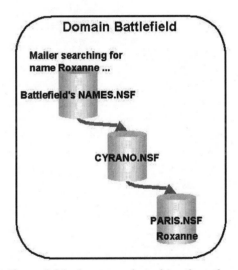

Figure 7.34   Sequence of searching through cascaded NABs

Instead of addressing mail to Roxanne Smith/Executives/Paris @ Paris or Roxanne Smith/Executives/Paris @ Paris @ Cyrano, for example, Christian would have to type only Roxanne Smith/Executives/Paris to send her a message.

Cascading NABs have the following major limitations:

- Notes will consult only the first address book listed in the NAMES= line for group, server, connection, and mail-in database documents.

- Notes will consult only the first address book listed in the NAMES= line for ambiguous user names or multiple users with the same name, if you do not type out the complete distinguished name.

EXAM TIP

Cascading Address Books change significantly in R4.5X, R4.6X, and R5. These releases address the limitations in cascading and improve the functions of cascading. R5 will also use a different name for cascading. The exam is based on the data as it is taken from Notes R4 through Notes R4.1X.

Merging and Splitting Domains

Domains can be merged or split. You might want to merge multiple domains in the following situations:

- When one company that uses Notes buys another company that uses Notes, and the company decides to place all the users in a single NAB.

- When a company that originally installed Notes using multiple domains within its different departments decides that combining these into a single domain makes more sense.

You might want to split a single domain into multiple domains in the following situations:

- When one company grows too large for a single NAB. (When a NAB grows too large, the performance decreases.)

- When a department or group needs to test Notes applications in a domain that is separate from the production domain.

- When one company that uses Notes buys another company that does not use Notes, and the company decides to place all users in separate NABs to maintain organizational structure.

When an administrator plans to merge a domain, she must consider and document all the places that the domain is referenced in the configuration of the Notes environment. Some of these locations include documents in the Public and Personal NABs, as well as in many of the databases in the environment. Look for references to the domain in all of these places, including the following specific areas:

- Server documents
- Connection documents
- Domain documents
- Person documents
- Mail-in database documents
- LocalDomainServers and OtherDomainServers group documents
- Location documents
- CERTLOG.NSF
- EVENTS4.NSF
- ADMIN4.NSF
- STATREP.NSF
- DOMAIN = line in the NOTES.INI (servers and workstations)

References to the domain will each have to be corrected and standardized to merge the domains. After making a list of the locations where the domain name will need to be changed, use the following as a checklist of steps necessary to merge two domains:

1. Decide which domain name will remain and which will be renamed. For ease of use, we will call these the *winning* domain and the *losing* domain.

2. Make the list of documents that need to be edited, as described earlier.

3. Bring down all the servers and clients in both domains to make a complete backup of the NAB for both domains. Then bring the servers back up. (This activity is called cycling the servers.)

4. In the backup copy of the NAB for the losing domain, change all the necessary documents to show the new domain name.

5. Remove any documents, such as Connection and domain documents, that may no longer be necessary.

6. Copy and paste the documents that you modified from the losing domain's NAB into the winning domain's NAB. Make replicas of the winning domain's NAB available on all servers in both domains.

7. Change the DOMAIN= line in the losing domain's NOTES.INI files.

8. Cycle the servers and test the new NAB.

When an administrator plans to split one domain into multiple domains, the same types of documents will need to be edited. Because its NAB defines a domain, however, a new NAB that is *not* a replica of the NAB in the original domain will need to be created. All of the changes to documents in the NAB should be made to the new NAB.

1. After you back up the NAB, so that you have a safe place to return in case of problems, use File . . . Database . . . New Copy to create a copy of the NAB for your new domain. Use a different name for this copy, such as NEWNAB.NSF.

2. In the new copy, you will delete people, servers, groups, connections, and other documents that do not need to exist in the new domain.

3. You will change server, domain, mail-in database, and other documents to reflect the new domain's name.

4. Bring down all the servers and workstations that will be part of the new domain.

5. Edit the DOMAIN= line to reflect the new domain name in the NOTES.INI files on the servers and workstations.

6. Delete the NAMES.NSF files that refer to the old domain.

7. Place replicas of the new NAB on each server in the new domain. Rename the NAB from NEWNAB.NSF to NAMES.NSF.

8. After modifying any custom applications and other databases that refer to the domain, bring the servers back up and test the new domains.

Questions

1. Company A has a new partner company, Company B. Both use separate domains and organizations. The employees of Company B need to use all of the servers and databases in Company A. The employees of Company A only need to use the server Server1/Houston/CompanyB. How should the administrators choose to cross-certify?

   a. Cross-certify the O of Company A with Server1 of Company B.

   b. Cross-certify the O of Company A with the O of Company B.

   c. Cross-certify the OU Houston of Company B with the O of Company A.

   d. Cross-certify each user of Company B with the O of Company A.

2. What level of cross certification is the most limiting but also the most secure?

   a. Server to O

   b. OU to OU

   c. OU to Server

   d. Server to Server

3. If Company A is hierarchical and Company B is flat, can they cross-certify?

   a. Yes

   b. No

4. When User1/Houston/Company A cross-certifies with Accounting/Company B, what happens to the user ID for User 1?

   a. It receives a new cross certificate certified by Company B.

   b. It receives a new cross certificate certified by Accounting/Company B.

   c. Nothing

   d. It generates a cross certificate to be stored in the Accounting/Company B ID file.

5. If User 1/Houston/Company A cross-certifies with Server1/ Accounting/Company B, where are the cross certificates stored?

    a. NAB in both Company A and Company B

    b. NAB in Company A and Server 1's Personal NAB in Company B

    c. User 1's Personal NAB and Server 1's Personal NAB

    d. User 1's Personal NAB and NAB in Company B

6. If User 1/Houston/Company A cross-certifies with Server1/ Accounting/Company B, who can use what?

    a. The users of Server 1 can use anything certified by Houston/ Company A.

    b. The users of Server 1 can use anything certified by Company A.

    c. User 1 can use Server1.

    d. User 1 can use anything certified by Server 1.

7. If Company A and Company B both cross-certify with each other at the O level, does that mean that any user from Company A can use any server in Company B?

    a. Yes

    b. No

8. If Company A is hierarchical and Company B is flat, who needs what kind of flat certificates required for communication?

    a. Company A, generated by Company B

    b. Company B, generated by Company A

    c. Company B, generated by Company B

    d. Company A, generated by Company A

9. Company A and Company B cross-certified at the OU level by sending safe copies of their OU certifier IDs to each other on disks. What can Company B do with your OU certifier?

    a. Create a cross-certificate with Company A.

    b. Certify users in Company B to use any items certified by the OU of Company A.

    c. Nothing

    d. Create a cross certificate with the OU of Company A.

10. If Walter in CompanyA@ACME tries to send a message to Krickett in CompanyB@Coyote, how must he address the message?

    a. Krickett@Coyote

    b. Krickett@CompanyB@Coyote

    c. Krickett@CompanyB@Coyote@ACME

    d. Krickett@Coyote@ACME

11. The users in DomainA need to send mail to DomainC. Messages to DomainC get routed through DomainB. What type of document is required in DomainA's address book to enable this?

    a. Adjacent domain document to DomainC

    b. Adjacent domain document to DomainB

    c. Nonadjacent domain document to DomainC

    d. Nonadjacent domain document to DomainB

12. The administrator of the company ACME wants to make sending mail messages to the company Coyote easier for his users. Right now, they have to know the full hierarchical name and domain of the users at Coyote to send mail to them. What can he do to make sending mail easier?

    a. Create a cascading NAB, with both ACME's NAB and COYOTE's NAB available to his users.

    b. Create an adjacent domain document for COYOTE.

    c. Create a mail routing Connection document for a server in COYOTE.

    d. Create a nonadjacent domain document for COYOTE.

13. If Elizabeth has created a cascading address book using her own company's NAMES.NSF and ACME.NSF from the ACME domain, and COYOTE.NSF from the COYOTE domain, where did she put these file names so that her users and servers know that they have these additional address books available? Where did she put the address books themselves?

   a. She put the file names in her server documents on all servers and put replicas of each address book on all servers.

   b. She put the file names in her server documents on all servers and put a replica of each address book on her hub server.

   c. She put the file names in her hub server's NOTES.INI and put a replica of each address book on all servers.

   d. She put the file names in all servers' NOTES.INIs and put replicas of each address book on all servers.

14. Both Domain C and Domain D use Domain B to route messages to Domain A. The users in Domain C address letters to users in Domain A like this: *Joe@DomainA*. The users in Domain D try this and get all their messages returned. What is different?

15. The administrators of Domain B have decided that Domain C routes too many messages to Domain A, and they can no longer pass those messages. The users of Domain D, however, are only passing three or four messages a day, which is fine. What can the administrators of Domain B do to continue to allow the users of Domain D to send mail through, while preventing the users of Domain C from sending mail?

16. Domain C and Domain D are merging to form a new Domain AB. What does the administrator of Server 1 in Domain B need to do to her server?

   a. Change the server document for Server 1.

   b. Change the NOTES.INI DOMAIN= line on Server 1.

   c. Delete the NOTES.INI file and allow it to be recreated.

   d. Delete the NAB and get a new one from Domain AB.

Answers

1. a. Cross-certify the O of Company A with Server1 of Company B.

2. d. Server to server

3. b. No

4. c. Nothing

5. d. User 1's Personal NAB and NAB in Company B

6. c. User 1 can use Server1.

7. b. No. Each server in Company B can restrict users from Company A by using server access lists or restrictions.

8. All of the above. Company B has its own flat certificates. Company A will have to generate flat certificates to give to Company B. Company A will have to obtain Company B's flat certificates.

9. d. Create a cross certificate with the OU of Company A.

10. a. Krickett@Coyote

11. b and c. Adjacent domain document to DomainB and a non-adjacent domain document to DomainC.

12. a. Create a cascading NAB, with both ACME's NAB and COYOTE's NAB available to his users. (For mail to route at all, an adjacent domain document and mail routing Connection document already had to exist.)

13. d. She put the file names in all servers' NOTES.INIs and put replicas of each address book on all servers.

14. The administrator of Domain C created a nonadjacent domain document for Domain A, so users would not have to use explicit names to route. The administrator of Domain D did not.

15. In the adjacent domain document from Domain B to Domain A, the administrators of Domain B can choose to Deny Mail from Domain C.

16. a and b. Change the server document for Server 1 and change the NOTES.INI DOMAIN= line on Server 1.

CHAPTER 8

Server Monitoring and Statistics

Throughout the previous chapters, you created and configured a Notes environment. You registered, installed, and configured servers and workstations, configured mail routing, and used replication schedules to keep your data updated. The environment you built is complex and requires a great deal of monitoring to ensure that it continues to function the way you expect—and that when it does not function correctly, you can troubleshoot the problem. Various types of tools are available for monitoring your Notes environment, including the following items:

- Notes Log (LOG.NSF) to track server-specific activities

- Notes Log Analysis (LOGA4.NSF) to search and analyze the Notes Log

- The Statistics and Events database (EVENTS4.NSF) to configure the monitoring on your server

- The Statistics database (STATREP.NSF) to monitor the statistics and events that occur on your server

- The Database Analysis database (DBA4.NSF) to monitor and analyze changes and events occurring in your important databases

- Server mailbox (MAIL.BOX) to monitor dead and pending mail (discussed in Chapter 4, "Notes Mail")

These databases, and the server tasks and tools that configure them and write to them, are the Notes administrator's main tools for monitoring and troubleshooting a Notes environment. This chapter describes many of the tools that Notes provides to assist you with monitoring your Notes environment.

Objectives

After reading this chapter, you should be able to answer questions based on the following objectives:

- Use the Notes Log to monitor events in your Notes environment.
- Configure and use the Log analysis to monitor and troubleshoot events in your Notes environment.
- Ensure that the proper server tasks are running to monitor your Notes environment.
- Use the Statistics reporting database to monitor the statistics of your Notes environment.
- Use the Events database to configure the monitoring of events in your Notes environment.
- Configure certain events in your Notes environment to trigger e-mail or other reactions.
- Troubleshoot and monitor databases using database analysis tools.

Notes Log

The Notes Log database is used to record all activity that occurs on a server. A server process called LOGGER monitors all activities and writes them to the log file. The file that stores this information is called LOG.NSF. This file is created automatically when a Notes server is first started. You can open the Notes Log in one of two ways:

- Choose File . . . Database . . . Open and select the server whose Log you want to view. In the database list, choose the Notes Log database.
- Choose File . . . Tools . . . Server Administration. In the Server Administration panel, click the System Databases button and choose Open Log.

Either method opens the Notes Log database and shows the list of available views in the database. The Notes Log displays information related to the following types of server activity:

- Mail routing
- Replication
- User sessions
- Database size and use
- Phone calls made or received by the server
- Other miscellaneous events

Each of the views in the Notes Log database presents an overview of the information. To see more detailed information, double-click a document in the view to open it. Some of the most useful views contained in the Notes Log and their uses are described next:

- *Database Sizes view* lists the database size, percentage used, and minutes of weekly usage. This feature can be used to monitor which databases in your environment have high or low use, as well as to check unused space in a database. A large amount of unused space indicates that you should run the server task COMPACT on that database. An example of the Database sizes view is shown in Figure 8.1.

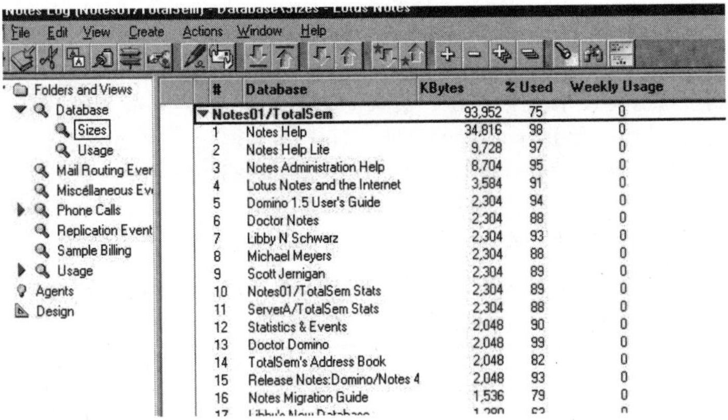

#	Database	KBytes	% Used	Weekly Usage
	▼ Notes01/TotalSem	93,952	75	0
1	Notes Help	34,816	98	0
2	Notes Help Lite	9,728	97	0
3	Notes Administration Help	8,704	95	0
4	Lotus Notes and the Internet	3,584	91	0
5	Domino 1.5 User's Guide	2,304	94	0
6	Doctor Notes	2,304	88	0
7	Libby N Schwarz	2,304	93	0
8	Michael Meyers	2,304	88	0
9	Scott Jernigan	2,304	89	0
10	Notes01/TotalSem Stats	2,304	89	0
11	ServerA/TotalSem Stats	2,304	88	0
12	Statistics & Events	2,048	90	0
13	Doctor Domino	2,048	99	0
14	TotalSem's Address Book	2,048	82	0
15	Release Notes:Domino/Notes 4	2,048	93	0
16	Notes Migration Guide	1,536	79	0
17	Libby's New Database	1,280	63	0

Figure 8.1 Database Sizes view

- *Database Usage view* lists the databases, the times, and dates the database was in use, and the user who activated the database. This view can enable you to verify database usage and can be used to help troubleshoot and verify replication. An example of the Database Usage view is shown in Figure 8.2.

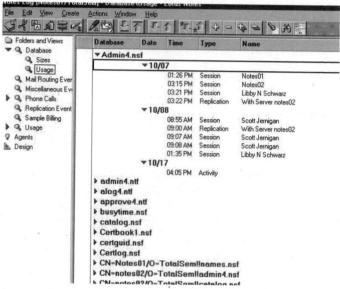

Figure 8.2   Database Usage view

- *Mail Routing Events view* shows in the view only the times the router was running. The documents in this view, however, show details about mail routing and other activities the router performed. This feature can be used to troubleshoot mail routing. The Mail Routing Events view is shown in Figure 8.3. A Mail Routing Events document is shown in Figure 8.4.

- *Miscellaneous Events view* shows only dates and times in the view. The documents, however, give details about the server sessions that do not appear in other views, including information on corrupted documents, some mail routing information, modem information (if applicable), and other events. This view is the first view to check when monitoring for server crashes and down time, corrupt databases, views, or documents, and user registration or session events. A Miscellaneous Events document is shown in Figure 8.5.

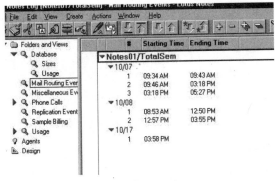

Figure 8.3   Mail Routing Events view

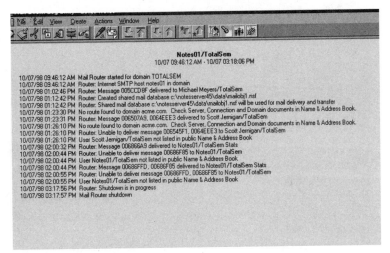

Figure 8.4   Mail Routing Events document

- *Phone calls By Date or By User view* shows information about calls made or received by a server. This feature can be used to verify replication or mail routing done over a phone line, to check the users and servers that connect using a modem, and to troubleshoot connection, mail routing, replication, and Passthru issues that may be related to modems or calls.

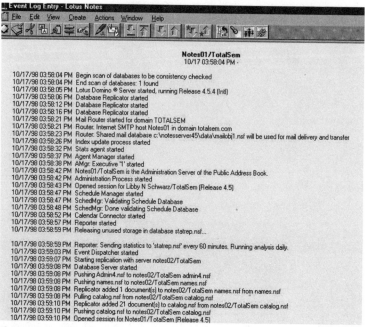

Figure 8.5 Miscellaneous Events document

- *Replication Events view* displays any replication events between servers, including information related to the time and date of the replication, how long the replication took to complete, and with which server the replications occurred. The documents in this view can help with troubleshooting server replication problems. Note that this view does not show workstation-to-server replication events or replications that failed without connecting to the remote server. The Replication Events view is shown in Figure 8.6. A Replication Events document is shown in Figure 8.7.

- *Usage By Date and By User views* show information related to the users who connected to the server, when and for how long they connected to the server, what databases they used while connected, and how many reads and writes they made while connected to the server. This view can help monitor database usage and monitor users who are active on the server and can help provide an overall view of the server use patterns. The Usage by Date view is shown in Figure 8.8.

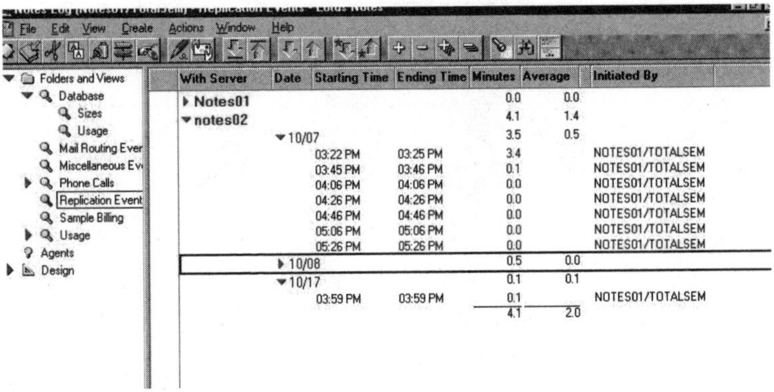

With Server	Date	Starting Time	Ending Time	Minutes	Average	Initiated By
▶ Notes01				0.0	0.0	
▼ notes02				4.1	1.4	
	▼ 10/07			3.5	0.5	
		03:22 PM	03:25 PM	3.4		NOTES01/TOTALSEM
		03:45 PM	03:46 PM	0.1		NOTES01/TOTALSEM
		04:06 PM	04:06 PM	0.0		NOTES01/TOTALSEM
		04:26 PM	04:26 PM	0.0		NOTES01/TOTALSEM
		04:46 PM	04:46 PM	0.0		NOTES01/TOTALSEM
		05:06 PM	05:06 PM	0.0		NOTES01/TOTALSEM
		05:26 PM	05:26 PM	0.0		NOTES01/TOTALSEM
	▶ 10/08			0.5	0.0	
	▼ 10/17			0.1	0.1	
		03:59 PM	03:59 PM	0.1		NOTES01/TOTALSEM
				4.1	2.0	

Figure 8.6   Replication Events view

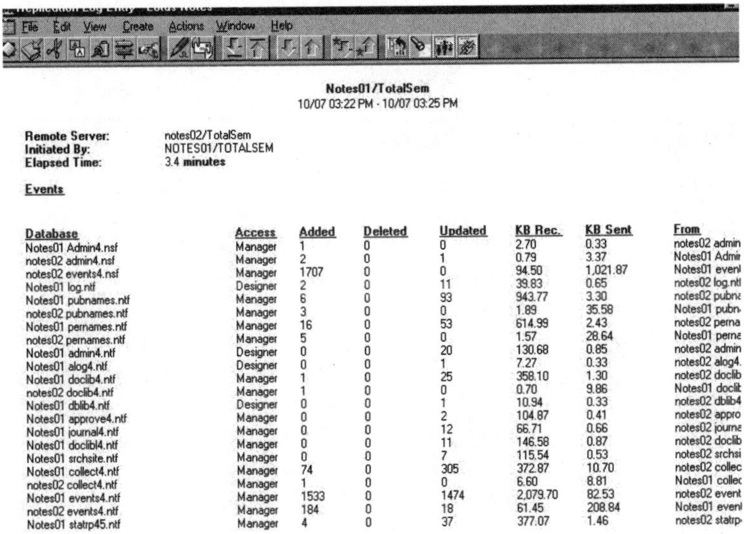

Notes01/TotalSem
10/07 03:22 PM - 10/07 03:25 PM

Remote Server: notes02/TotalSem
Initiated By: NOTES01/TOTALSEM
Elapsed Time: 3.4 minutes

Events

Database	Access	Added	Deleted	Updated	KB Rec.	KB Sent	From
Notes01 Admin4.nsf	Manager	1	0	0	2.70	0.33	notes02 admin
notes02 admin4.nsf	Manager	2	0	1	0.79	3.37	Notes01 Admi
notes02 events4.nsf	Manager	1707	0	0	94.50	1,021.87	Notes01 event
Notes01 log.ntf	Designer	2	0	11	39.83	0.65	notes02 log.ntf
Notes01 pubnames.ntf	Manager	6	0	93	943.77	3.30	notes02 pubna
notes02 pubnames.ntf	Manager	3	0	0	1.89	35.58	Notes01 pubn
Notes01 pernames.ntf	Manager	16	0	53	614.99	2.43	notes02 perna
notes02 pernames.ntf	Manager	5	0	0	1.57	28.64	Notes01 perna
Notes01 admin4.ntf	Designer	0	0	20	130.68	0.85	notes02 admin
Notes01 alog4.ntf	Designer	0	0	1	7.27	0.33	notes02 alog4.
Notes01 doclib4.ntf	Manager	1	0	25	358.10	1.30	notes02 doclib
notes02 doclib4.ntf	Manager	1	0	0	0.70	9.86	Notes01 doclit
Notes01 dblib4.ntf	Designer	0	0	1	10.94	0.33	notes02 dblib4
Notes01 approve4.ntf	Manager	0	0	2	104.87	0.41	notes02 appro
Notes01 journal4.ntf	Manager	0	0	12	66.71	0.66	notes02 journa
Notes01 doclibl4.ntf	Manager	0	0	11	146.58	0.87	notes02 doclib
Notes01 srchsite.ntf	Manager	0	0	7	115.54	0.53	notes02 srchsi
Notes01 collect4.ntf	Manager	74	0	305	372.87	10.70	notes02 collec
notes02 collect4.ntf	Manager	1	0	0	6.60	8.81	Notes01 collec
Notes01 events4.ntf	Manager	1533	0	1474	2,079.70	82.53	notes02 event
notes02 events4.ntf	Manager	184	0	18	61.45	208.84	Notes01 event
Notes01 statrp45.ntf	Manager	4	0	37	377.07	1.46	notes02 statrp

Figure 8.7   Replication Events document

Some settings in the NOTES.INI file can affect what gets written to the Notes Log and how long data remains in the Notes Log. The Notes Log can grow to be a large and unwieldy database without some of the settings described in Table 8.1. This table provides only enough information to show the available options; therefore, for more information on NOTES.INI settings and the Notes Log, see the Notes Administration Help database.

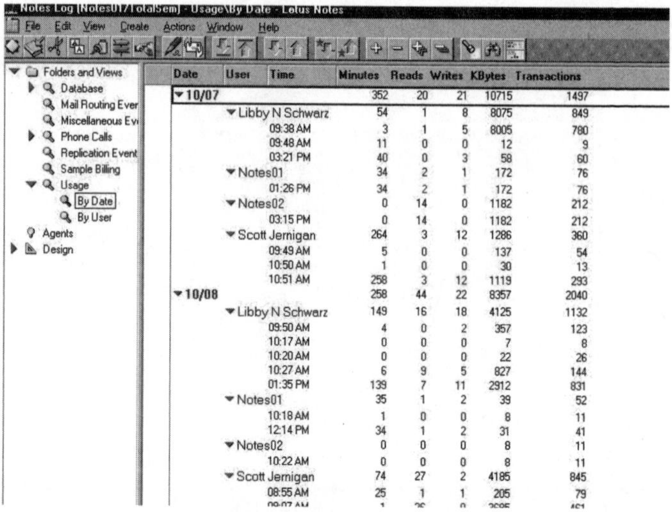

Notes Log [Notes01/TotalSem] - Usage\By Date - Lotus Notes

File Edit View Create Actions Window Help

	Date	User	Time	Minutes	Reads	Writes	KBytes	Transactions
Folders and Views								
Database	▼ 10/07			352	20	21	10715	1497
Mail Routing Ever		▼ Libby N Schwarz		54	1	8	8075	849
Miscellaneous Ev			09:38 AM	3	1	5	8005	780
Phone Calls			09:48 AM	11	0	0	12	9
Replication Event			03:21 PM	40	0	3	58	60
Sample Billing		▼ Notes01		34	2	1	172	76
Usage			01:26 PM	34	2	1	172	76
By Date		▼ Notes02		0	14	0	1182	212
By User			03:15 PM	0	14	0	1182	212
Agents		▼ Scott Jernigan		264	3	12	1286	360
Design			09:49 AM	5	0	0	137	54
			10:50 AM	1	0	0	30	13
			10:51 AM	258	3	12	1119	293
	▼ 10/08			258	44	22	8357	2040
		▼ Libby N Schwarz		149	16	18	4125	1132
			09:50 AM	4	0	2	357	123
			10:17 AM	0	0	0	7	8
			10:20 AM	0	0	0	22	26
			10:27 AM	6	9	5	827	144
			01:35 PM	139	7	11	2912	831
		▼ Notes01		35	1	2	39	52
			10:18 AM	1	0	0	8	11
			12:14 PM	34	1	2	31	41
		▼ Notes02		0	0	0	8	11
			10:22 AM	0	0	0	8	11
		▼ Scott Jernigan		74	27	2	4185	845
			08:55 AM	25	1	1	205	79

Figure 8.8   Usage by Date view

Analyzing the Notes Log

Because the Notes Log monitors all server activity, it grows very large very quickly. When you are researching or troubleshooting a particular problem or topic, it may be slow and inconvenient to search the Notes Log manually. To make searches in the Notes Log faster and more efficient, Notes provides a Log Analysis tool and database. To use the Log Analysis tool, access the Server Administration panel by choosing File . . . Tools . . . Server Administration from the menus. In the Server Administration panel, select the Servers button and click Log Analysis . . . . The Server Log Analysis dialog box is displayed, as shown in Figure 8.9.

This dialog box enables you to choose where the results of the Log Analysis will be written. By default, they will be written to a database called LOGA4.NSF on the local machine that is running the analysis. You can change these settings by clicking the Results Database button, as shown in Figure 8.10. The Results Database dialog box also enables you to select whether this analysis will be appended, or added, to the current information in the Log Analysis database, or whether the current analysis will overwrite any information in the Log Analysis database.

Table 8.1 NOTES.INI Settings That Affect the Notes Log

Setting and Syntax	Parameters	Use and Other Information
LOG= logfilename, log_option, not_used, days, size	■ The logfilename parameter enables you to verify the name of the log file, usually LOG.NSF. ■ The log_option parameter enables you to add options to the logging process, such as adding a database FIXUP and logging to the console. ■ The not_used parameter is not used. ■ The days parameter enables you to determine how many days to retain log documents. The default is seven days. ■ The size parameter enables you to set the amount of text allowed in log documents.	This setting is always in the NOTES.INI. You may want to edit it to change the value of the days or size parameter. An example of the LOG setting that would change the name of the log file and limit the number of days to three would be the following: LOG=noteslog.nsf, 1, 0, 3, 40000

continues

Table 8.1 Continued.

Setting and Syntax	Parameters	Use and Other Information
Log_MailRouting= value	The value parameter can be set at 10, 20, 30, or 40. The default is 20. A setting of 10 writes only errors in mail routing to the Notes Log. A setting of 40 uses verbose messages.	If you find that you are not receiving enough mail routing information to troubleshoot a problem, you may increase the value. Similarly, if mail routing is occurring normally and your Notes Log size needs to be reduced, you may choose to lower the value. This setting can be edited using a server configuration document.
Log_Replication = value	The value parameter can be set at 0, 1, 2, 3, 4, or 5. There is no default value. A setting of 0 does not log replication events.	This option can be enabled on the Advanced Setup tab during server configuration. You can also add the option using a server configuration document or directly to the NOTES.INI file.
Log_Sessions = value	The value parameter can be set at 0 or 1. There is no default value. A value of 0 will not log user session events. A value of 1 will log all user session events.	This option can be enabled on the Advanced Setup tab during server configuration. You can also add the option using a server configuration document or directly to the NOTES.INI file.

Setting and Syntax	Parameters	Use and Other Information
Log_Tasks = value	The value parameter can be set at 0 or 1. There is no default value. A value of 0 will not send status information about server tasks to the console and the Log. A value of 1 will send the status of server tasks both to the Log and to the console.	You can add the option using a server configuration document or directly to the NOTES.INI file.
Mail_Log_To_MiscEvents = value	The value parameter can be set at 0 or 1. There is no default value. A value of 0 will not display mail events in the Miscellaneous Events view. A value of 1 will display mail events in both the Miscellaneous Events view and the Mail Routing Events view.	You can add the option directly to the NOTES.INI file. If the option is not in the NOTES.INI file, mail routing events will not be written the Miscellaneous Events view.

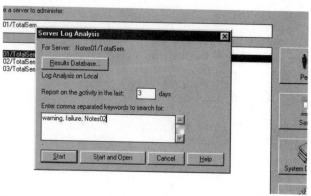

Figure 8.9   Server Log Analysis dialog box

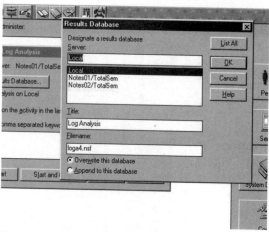

Figure 8.10   Selecting a Results Database

In the Server Log Analysis dialog box, select the number of days of logged activity that you want to search. The default is one day. This causes the analysis to look only at documents created within the last 24 hours. To create the search criteria, type comma-separated keywords in the search area. Do not use phrases; use only single words. Notes runs a search based on each of the keywords and returns all the documents that contain any of the words to the analysis database, as shown in Figure 8.11.

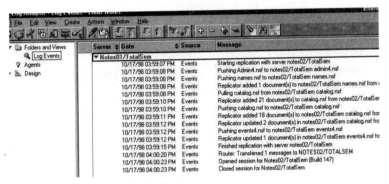

Figure 8.11 Log Analysis database

Statistics and Events

Notes uses statistics and events to monitor the activity on Notes servers. A *statistic* is a piece of information gathered regularly about a resource. This resource can be directly related to Notes, such as dead or pending mail, or it can be related to the general health of the server, such as disk space or memory. An *event* occurs when a particular statistic reaches a preset threshold, or when Notes network activities occur. The types of statistics that can be gathered fall into the following categories:

- *Session information*, which lists the date and time of the reports and the server location and administrator.

- *Disk statistics*, which list the amount and percentage of free space on each of the drive volumes.

- *Memory statistics*, which list the amount of memory that is free, allocated, and available.

- *Server configuration information*, which lists information regarding the type of processor, operating system, version of Notes, and path for the Notes data files.

- *Server load statistics*, which list the number of users, the number of transactions, the peaks in users and transactions, and the number of user sessions that are dropped.

- *Replication statistics*, which list the number of successful and failed replications, and the number of documents added, deleted, and updated by the replications.

- *Database statistics*, which describe buffer pool and NSF pool information.

- *Notes mail statistics*, which list the number of dead, pending, delivered, and transferred messages, including mail delivery times and sizes.

- *Mail gateway statistics*, which list information for any installed and available mail gateways, such as SMTP or cc:Mail gateways.

- *Communications and protocol statistics*, which give information on ports and protocols.

The events gathered and reported in Notes are grouped into the following categories:

- *Communication events* related to modem problems, local and wide area network problems, and port or protocol problems.

- *Mail events* are any mail-routing problems, such as routing failures or dead mail.

- *Replica events* are database replication problems, such as failed replications.

- *Resource events* are system resource problems, such as low memory or low disk space.

- *Security events* are related to server and database access and user and server IDs.

- *Server events* are related to server tasks that do not run or shut down unexpectedly.

- *Statistics events* are reported when certain statistics reach a threshold set by the administrator.

- *Miscellaneous events* are problems that occur that do not fall into any of the other categories.

The events that occur are also categorized by the level of severity, ranging from Normal to Fatal. These levels of severity are completely subjective and can be interpreted in different ways by the administrator:

- *Fatal events* should be interpreted as server crashes.

- *Failure events* should be interpreted as failures that enable the server to stay up but not function correctly, such as a NAMES.NSF no longer being available.

- *Warning (high) events* should be interpreted as failures that require some administrator intervention, such as routing or replication failures.

- *Warning (low) events* should be interpreted as events that signify poor performance of the server, such as memory or disk space being lower than optimal.

- *Normal events* should be interpreted as informational status messages.

Reporter and Event Server Tasks

Statistics and Events are gathered from the Notes server and written to the Statistics Reporting database (STATREP.NSF), based on configuration information stored in the Statistics and Events database (EVENTS4.SNF). This information is gathered using two Notes server tasks, called REPORTER and EVENT. The EVENT task monitors the statistics and events that are occurring on the server based on the information in the EVENTS4.NSF database. The REPORTER task then collects the statistics and events for the server and reports them in the appropriate location. This feature can report them to the statistics mail-in database, the Statistics Reporting database (STATREP.NSF), or it can mail them to a specified user.

Both server tasks, REPORTER and EVENT, can be loaded manually from the server console and can be added to the SERVERTASKS= line in the NOTES.INI, as shown in Figure 8.12.

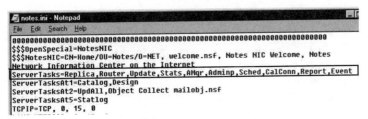

Figure 8.12 NOTES.INI SERVERTASKS= line

To load the tasks manually at the server console, use the following commands:

- LOAD REPORT
- LOAD EVENT

The first time the REPORTER task runs on the server, it completes the following tasks:

- Creates the Statistics Reporting database (STATREP.NSF), if it does not already exist.

- Creates the Statistics and Events database (EVENTS4.NSF), if it does not already exist.

- Creates a Mail-in database document in the NAB, pointing to the Statistics Mail-in database for the server.

- It opens both STATREP.NSF and EVENTS4.NSF for its use.

- Specifies the default collection interval for statistics in the Statistics and Events database (EVENTS4.NSF).

The first time the EVENT task runs on the server, it completes the following tasks:

- Creates the Statistics and Events database (EVENTS4.NSF), if it does not already exist.

- Creates the Statistics Reporting database (STATREP.NSF), if it does not already exist.

- It opens both STATREP.NSF and EVENTS4.NSF for its use.

- Copies the list of servers in the Server . . . Servers view of the NAB to the Servers to Monitor view in the Statistics and Events database (EVENTS4.NSF).

Statistics Reporting database

The Statistics Reporting database (STATREP.NSF) displays the statistics and events gathered for a server, or a group of servers. You can use STATREP.NSF to gather statistics for multiple servers in an environment, if you wish. The views in the Statistics Reporting database are described and discussed next:

- System Statistics Reports view shows information about the statistic collection time, the space available, the amount of dead or pending mail, the number of users, and the transactions per minute, as shown in Figure 8.13.

- Mail and database statistics reports view shows information about dead, pending, and routed mail, as well as failed and successful replications, as shown in Figure 8.14.

Collection Time	Space on Data Path	Swap File/Sysvol Size	Dead Mail	Pend. Mail	Users	Mem Alloc
Notes01/TotalSem						
10/08 02:58 PM	1,375,764,480	NA	0	0	1	6,382,
10/08 01:58 PM	1,417,871,360	NA	0	0	1	5,714,
10/08 11:58 AM	1,371,930,624	NA	2	0	1	8,965,
10/08 10:58 AM	1,371,996,160	NA	2	0	0	7,719,
10/08 09:58 AM	1,376,518,144	NA	2	0	1	6,982,
10/08 08:58 AM	1,410,465,792	NA	2	0	1	4,286,
10/07 05:07 PM	1,386,020,864	NA	2	0	1	6,745,
10/07 04:07 PM	1,382,318,080	NA	2	0	1	6,243,
10/07 02:55 PM	1,382,252,544	NA	2	0	2	6,473,
10/07 01:55 PM	1,389,133,824	NA	1	0	2	6,106,

Figure 8.13 System statistics view

Collection Time	Dead Mail	Mail Routed	Pend. Mail	Failed Replications	Successful Replications
Notes01/TotalSem					
10/08 02:58 PM	0	1	0	N/A	336
10/08 01:58 PM	0	1	0	N/A	168
10/08 11:58 AM	2	3	0	N/A	505
10/08 10:58 AM	2	3	0	N/A	337
10/08 09:58 AM	2	N/A	0	N/A	169
10/08 08:58 AM	2	N/A	0	N/A	N/A
10/07 05:07 PM	2	N/A	0	N/A	340
10/07 04:07 PM	2	N/A	0	N/A	172
10/07 02:55 PM	2	4	0	N/A	N/A
10/07 01:55 PM	1	2	0	N/A	N/A

Figure 8.14 Mail and Database statistics view

- Communications statistics view shows any port errors or retransmissions that may have occurred.

- Network statistics view shows the enabled network ports (protocols).

- Alarms view shows information about any alarms that have occurred. An *alarm* occurs when a statistic crosses a preset threshold, as shown in Figure 8.15.

- Events view shows events that have occurred based on the definitions of events in the Statistics and Reporting database (EVENTS4.NSF).

- Spreadsheet export view shows statistics and information gathered from other views, organized in a useful view for exporting. To export data to a spreadsheet, choose File . . . Export, and select the type of file you want to create.

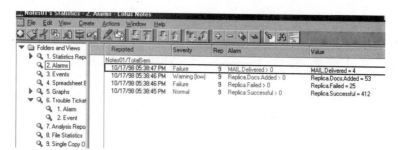

Figure 8.15 Alarm view in the Statistics and Reporting database

■ Graph views for System statistics, load, and resources, show the pertinent information in a somewhat graphical form. As you can see in Figure 8.16, it is only a limited graphical representation of the information.

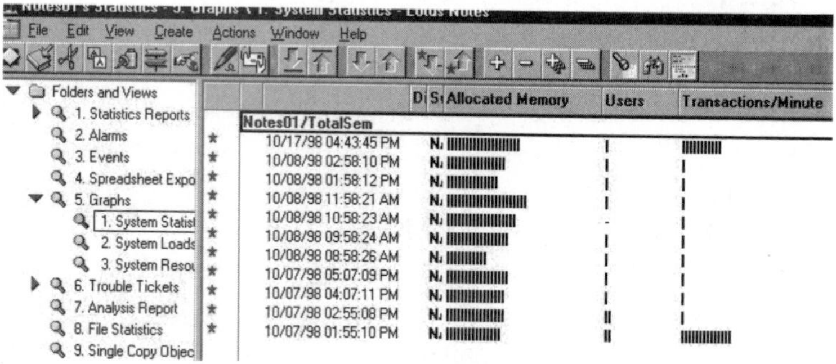

Figure 8.16 System statistics graph view

■ Alarm and Event trouble tickets views show the trouble tickets that have been created for the events and alarms that have occurred in the environment. Trouble tickets are discussed later in this chapter.

■ Analysis report view contains documents that show a low, high, and average for specified statistics.

■ File statistics view shows information about all the databases stored on the server, including filename, database title, replica ID, file size and free space, percentage used, and the reads, writes, and usage. This view is shown in Figure 8.17.

Folders and Views	Server	Filename	DB Title	Replica ID	File Size
▶ 🔍 1. Statistics Repor					
🔍 2. Alarms	★ Notes01/TotalSem	THE SCOTTISH HISTORICAL COOKBOOK.NSF	The Scottish Historical Cookbook	8625666E:006ACB95	262,144
🔍 3. Events					
🔍 4. Spreadsheet E>	★ Notes01/TotalSem	STATS730.NSF Inherit Design From: StdR45Mail	Servera/Totalsem Stats	86256667:0074A435	2,359,296
▶ 🔍 5. Graphs					
▼ 🔍 6. Trouble Tickets					
🔍 1. Alarm	★ Notes01/TotalSem	STATS288.NSF Inherit Design From: StdR45Mail	Notes01/Totalsem Stats	86256696:005013DF	2,359,296
🔍 2. Event					
🔍 7. Analysis Report	★ Notes01/TotalSem	STATRP45.NTF Design Template: StdR4StatReport	Statistics Reporting	85256373:005FE750	1,048,576
🔍 8. File Statistics					
🔍 9. Single Copy Ob	★ Notes01/TotalSem	STATREP.NSF Inherit Design From: StdR4StatReport	Notes01'S Statistics	80256696:042E061D	1,048,576
🖗 Agents					
▶ 📖 Design	★ Notes01/TotalSem	STATREP.NTF	Notes01'S Statistics	80256696:042E061D	1,048,576
	★ Notes01/TotalSem	SRCHSITE.NTF Design Template: StdNotesSearchSite	Search Site	8525632D:0055EA6C	524,288
	★ Notes01/TotalSem	RESRC45.NTF Design Template: Std45ResourceRese	Resource Reservations (4.5)	8525637D:00812B9B	524,288
	★ Notes01/TotalSem	PUBWEB45.NTF Design Template: StdR45WebNavigat	Server Web Navigator 4.5	852563AA:004488 1A	2,097,152
	★ Notes01/TotalSem	PUBNAMES.NTF Design Template: StdR4PublicAddress	Public Address Book	852560CF:00644F08	1,572,864
	★ Notes01/TotalSem	PERWEB45.NTF Design Template: StdR45PersonalWet	Personal Web Navigator 4.5	85256345:0066F62C	1,048,576
	★ Notes01/TotalSem	PERNAMES.NTF Design Template:	Personal Address Book	852560CF:006DA0FD	1,048,576

Figure 8.17 File statistics view

- Single Copy Object Store view contains statistics that relate to Shared mail and the Shared mail database. Use this view to keep track of the Shared mail database size and number of messages that are using Shared mail, if you have enabled Shared mail in your environment.

Use the Statistics Reporting database to monitor the statistics and events for your server or servers. You can also use the Statistics Reporting database to create trouble tickets for alarms and events that have been reported. A *trouble ticket* is a document that can assign a particular user or group of users to look into a specific event or alarm. The trouble ticket is created manually in the Statistics Reporting database and then either sent to the specified user or saved in the Trouble Tickets views in the database. To create a trouble ticket for an event or alarm, use the following steps:

1. In the Alarms or Events view, select the Alarm or Event for which you wish to create a trouble ticket.

2. Choose Create . . . Alarm Trouble Ticket or Create . . . Event Trouble Ticket from the menus.

3. A Trouble Ticket document will be created based on the information contained in the selected event or alarm, as shown in Figure 8.18.

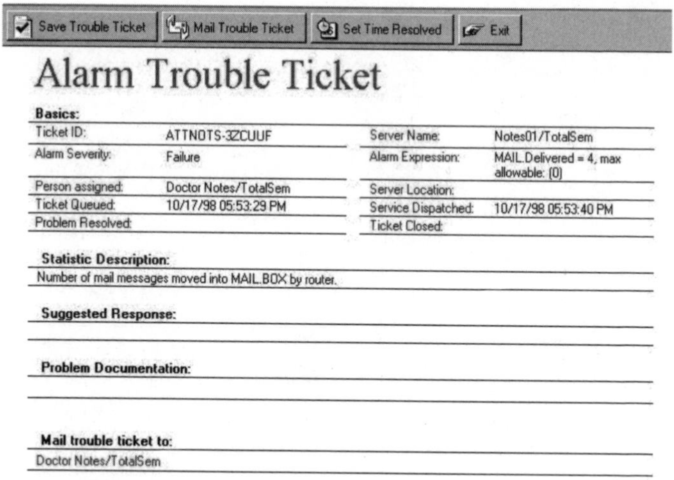

| Save Trouble Ticket | Mail Trouble Ticket | Set Time Resolved | Exit |

Alarm Trouble Ticket

Basics:

Ticket ID:	ATTNOTS-3ZCUUF	Server Name:	Notes01/TotalSem
Alarm Severity:	Failure	Alarm Expression:	MAIL.Delivered = 4, max allowable: (0)
Person assigned:	Doctor Notes/TotalSem	Server Location:	
Ticket Queued:	10/17/98 05:53:29 PM	Service Dispatched:	10/17/98 05:53:40 PM
Problem Resolved:		Ticket Closed:	

Statistic Description:
Number of mail messages moved into MAIL.BOX by router.

Suggested Response:

Problem Documentation:

Mail trouble ticket to:
Doctor Notes/TotalSem

Figure 8.18   Alarm Trouble Ticket document

4. Complete the trouble ticket by assigning a user or a group of users to be responsible for the event or alarm. Save or send and save the document.

Statistics and Events Database

The Statistics and Events database (EVENTS4.NSF) is used to configure the monitoring activities on the server. The database can be used to define a specific event to be reported, such as a change in the ACL or a failed replication. The database can also define the thresholds for statistics, and it can request statistic information from other servers. The views in the database enable the administrator to configure these options:

- The Database monitors view allows an administrator to create ACL and Replication monitors. The Database monitors view is shown in Figure 8.19.

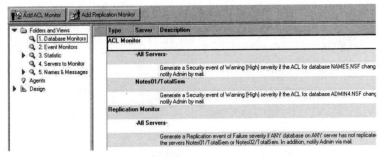

Figure 8.19 Database monitors view

To create a database monitor, select either the Add ACL monitor or Add Replication monitor action buttons. For either a Replication or an ACL monitor, complete information on which database and which server you want to monitor, what level event to create if the threshold is reached, and what to do if the event is triggered, as shown in Figures 8.20 and 8.21.

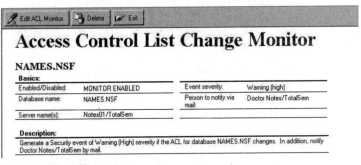

Figure 8.20 ACL monitor

- The Event monitors view enables an administrator to define what to do when certain types of events occur. If an event that falls into the Mail category occurs, for example, and is listed as Failure severity, an event monitor document can log this event to a database, send the event to another server, or send the event to a user or group. The Event monitors view is shown in Figure 8.22.

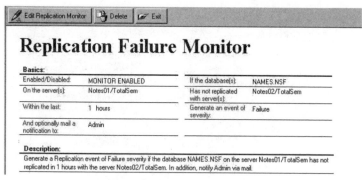

Figure 8.21 Replication monitor

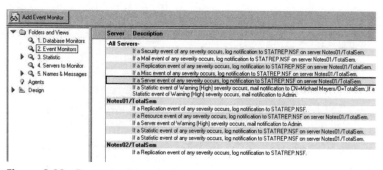

Figure 8.22 Event monitors view

- The Statistic monitors view enables an administrator to define what to do when the specified monitors reach the specified thresholds. If the Number of Sessions dropped in mid-transaction statistic (Server.Sessions.Dropped) becomes greater than 10, for example, this document enables the administrator to generate a statistic event of a certain severity. The documents in the Event monitors view then determine what is done with that event. The Statistic monitors view is shown in Figure 8.23.

- The Request for Remote statistics view enables an administrator to send a request for statistics to any server in the environment. The request can show a particular statistic, or if left blank, it can return all available statistics. The statistics can be sent to any mail file.

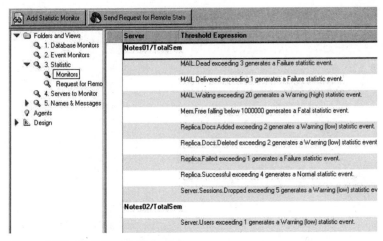

Figure 8.23   Statistic monitors view

- The Server to monitor view enables the administrator to define which servers should be monitored. The server-to-monitor document also enables the administrator to define how often the statistics should be gathered, how often the statistics should be analyzed, and how the statistics report should be stored (directly in STATREP.NSF or mailed to a mail-in database). An example of a server to monitor document is shown in Figure 8.24.

| | Save Server To Monitor | Delete | Exit |

Server to Monitor

Notes01/TotalSem

Basics:

Server name:	Notes01/TotalSem	Server administrators: Libby N Schwarz/TotalSem, Admin
Server title:		Report method: Log to Database
Domain name:	TotalSem	Enter server name: Notes01/TotalSem
		Database to receive statrep.nsf
		reports:
Collection interval in (minutes):	120	Analysis interval: ● Daily ○ Monthly ○ Weekly ○ Never
Server description:	Hub server for TotalSem	

Description:
Report statistics for server Notes01/TotalSem, to database 'statrep.nsf' on server Notes01/TotalSem.
Sample statistics every 120 minutes.
Analize statistics daily.

Figure 8.24   Server to monitor document

- The Messages view describes the message written to the Notes Log when an event occurs. The message is sorted based on the type and severity of the event.

- The Messages by text view sorts the messages written to the Notes Log in alphabetical order, rather than by event type.

- The Notification methods view displays the available methods of sending a notification of an event.

- The Statistic Names view describes all the available statistics and whether their value is text, time, or number.

- The Thresholds view shows all the statistics that have recommended thresholds.

The EVENTS4.NSF configures the statistics that should be gathered and reported and configures the information that should be gathered and reported about specific server events that may occur.

EXAM TIP

For the System Administration II exam, it is important to be able to distinguish between the REPORTER and EVENT tasks and between the Statistics Reporting database (STATREP.NSF) and the Statistics and Events database (EVENTS4.NSF).

Viewing Statistics at the Server Console

You can also view a specific statistic—or all statistics on demand—at the server console or at the remote console. To view a particular statistic, type in the following command at the console:

```
SHOW STAT statisticname
```

NOTE

To run this command, you must be running the server task STATS. This task can be loaded at the console by typing LOAD STATS or by adding STATS to the SERVERTASKS= line in the NOTES.INI file.

If you type the command without specifying a statistic, all statistics for the server will be displayed on the console. Unfortunately, because there are so many statistics available, this information scrolls by rather quickly. Figure 8.25 shows the results of typing SHOW STAT at the server console.

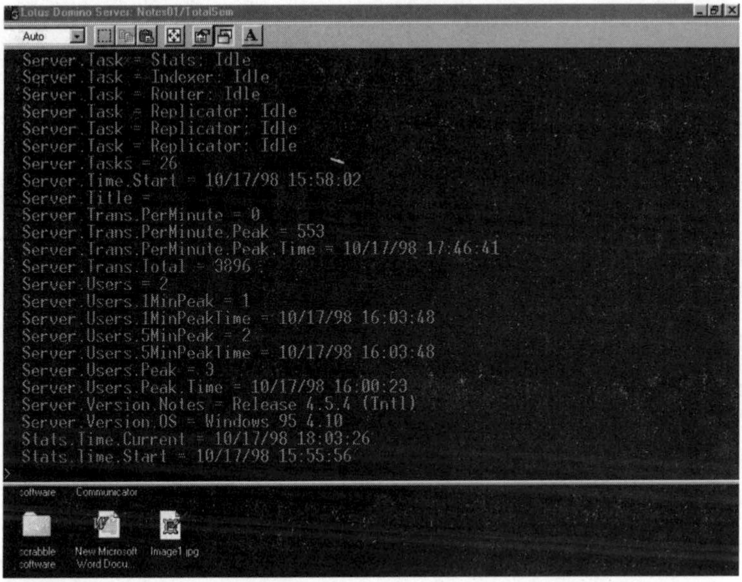

Figure 8.25 SHOW STAT results

If you type the command with a particular statistic name, that statistic will be displayed on the console, as shown in Figure 8.26. The following items are some examples of statistics you might want to see:

- SHOW STAT Disk.C.Free shows the free space on the C: drive.

- SHOW STAT Mem.Availability shows whether memory is plentiful, painful, or normal.

- SHOW STAT Server.Users shows the current number of users with open sessions.

For more information on available statistics, please refer to the Notes Administration Help database.

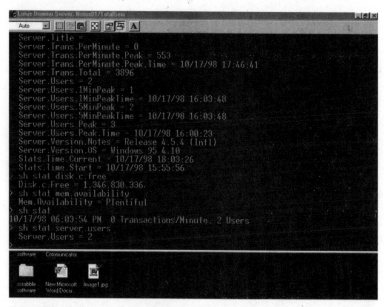

Figure 8.26 Statistics viewed on the console

Analyzing Databases

Some issues that all administrators monitor closely are disk space, database usage, and database performance. Much of the information that will enable an administrator to make a judgment about these issues can be found in the Notes Log and in the Statistics and Reporting database, as we have discussed throughout this chapter. Notes offers one additional tool for tracking information about databases, called Database Analysis.

A database analysis can tell an administrator about reads, writes, changes, and replications of a particular database. A database analysis is similar to a log analysis. We used a log analysis because even though all the information about the server activities was contained in the Notes Log, the Notes Log is too large to search and monitor effectively. Similarly, all the information necessary to monitor database activity is available in the Notes Log and the Statistics and Reporting databases. These databases are large, however, and it would be difficult to search through both for all the information necessary to monitor the activity for a particular database. The database analysis searches the Notes Log and the Statistics and Reporting databases for the information specified by the administrator and writes the results to a Results database, called DBA4.NSF, by default.

To analyze a database, open the Server Administration panel by choosing File . . . Tools . . . Server Administration from the menus. In the Server Administration panel, click the Database Tools icon. The dialog box shown in Figure 8.27 enables you to select a database to analyze. Then select the Analyze a Database tool from the drop-down Tools menu.

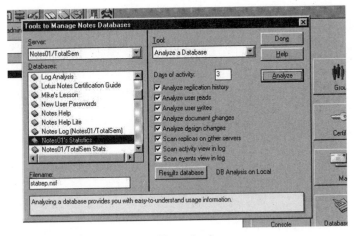

Figure 8.27 Tools to manage Notes databases

NOTE
The database tools available in the Server Administration panel and the method of selecting a tool change somewhat in appearance from Notes R4.0 to 4.5X or 4.6X. The functions described remain the same, however.

Select the number of days of activity that you want to analyze for the database. This number will depend on how often the database is used. The default number of days is one. Then consider the following choices and select the items that you want Notes to analyze:

- Analyze replication history
- Analyze user reads
- Analyze user writes
- Analyze document changes

- Analyze design changes
- Scan replicas on other servers
- Scan activity view in log
- Scan events view in log

After selecting the events and changes to analyze, click the Results Database button to specify the location and name of the Results database. By default, the database is stored on the local machine starting the analysis, with a filename of DBA4.NSF and database title of DB Analysis. The Results Database dialog box also enables you to choose whether to overwrite any other data already in the DB Analysis database, or whether to append, or add, the information to the database. Click Analyze to create the DB Analysis database, and add the icon for it to your workspace.

The DB Analysis database shows the requested information sorted in four views:

- By Date
- By Event Type
- By Source
- By Source Database

Figure 8.28 shows the By Date view. Figure 8.29 shows an example of a Database Analysis document.

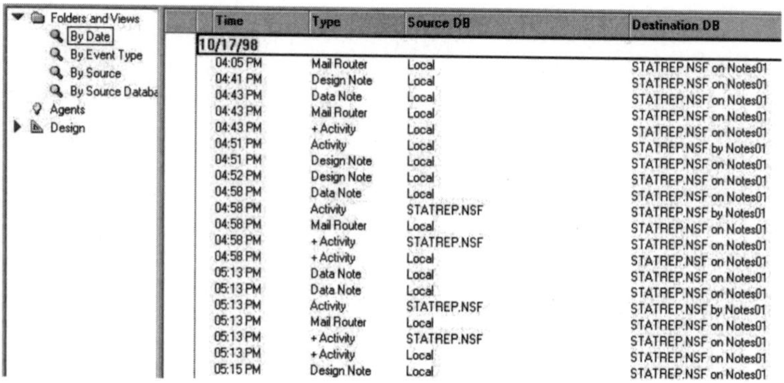

Figure 8.28   By Date view of the Database Analysis database

Database Analysis Results

Basics

Date:	10/17/98
Time:	05:38:48 PM
Source of Event Information:	Log File
Event Type:	+ Activity

Source

Source Database:	STATREP.NSF
Source:	Notes01

Destination

Destination Database:	
Destination Machine:	Notes01

Description

Description:	6 documents read from Source

Figure 8.29   Database Analysis document

Questions

1. Joe and Amy disagree about how to view the Notes Log file for Server 1. Joe says you choose File . . . Database . . . Open, select Server 1 as the server, and choose Notes Log (Server1). Amy says to use File . . . Tools . . . Server Administration . . . System Databases . . . Open Log. Who is right?

   a. Joe

   b. Amy

   c. Neither

   d. Both

2. Katie is trying to determine whether there are databases on her server that need to be compacted. How can she find this out?

   a. Open the Notes Log on her server and look at the Database Usage view

   b. Open the Database Catalog on her server and look at the Database Usage view

   c. Open the Notes Log on her server and look at the Database Sizes view

   d. Open the Database Catalog on her server and look at the Database Sizes view

3. Melinda is trying to check for replications from her server (Server 1) to another server (Server 2) that failed when the servers tried to connect to each other. Where can she look?

   a. Notes Log . . . Replication Events view

   b. Notes Log . . . Miscellaneous Events view

   c. Notes Log . . . Usage by Date view

   d. Notes Log . . . Database Usage view

4. Sal wants the Notes Log on Server 1 to remove documents older than five days, because the Notes Log is getting too big. How can he do this?

   a. `NOTESLOG= LOG.NSF,1,0,5,45000`

   b. Create a Log settings document in the NAB; set the delete period to five.

   c. `LOG = LOG.NSF, 1,0,5,45000`

   d. He does not need to do anything. This setting is the default.

5. Harry notices that an error occurs when Server2 tries to contact his server, Server1, for replication or mail routing. He wants to try to find out how long this error has been appearing. What might he do to find this out as easily as possible?

   a. Use a Log Analysis to search for the name Server2 in the Notes Log database.

   b. Use a Database Analysis to search for the name Server2 in the Notes Log database.

   c. Create a Log Analysis document in the NAB to search for the name Server2.

   d. Choose Edit . . . Find in the Notes Log to look for the name Server2.

6. Erna wants to receive a mail message whenever her server begins to run low on disk space. How can she configure her server to do this?

   a. Use the `STATREP.NSF` database to configure a Statistic Monitor to monitor disk space and an Event Monitor. Cause the Event monitor to mail a notification to her when the Statistic Monitor causes an Event.

   b. Use the `EVENTS4.NSF` database to configure a Statistic Monitor to monitor disk space and an Event Monitor. Cause the Event monitor to mail a notification to her when the Statistic Monitor causes an Event.

   c. Use the `STATREP.NSF` database to configure a Statistic Monitor that sends her a message when her server runs low on disk space.

   d. Use the `EVENTS4.NSF` to configure a Statistic Monitor that sends her a message when her server runs low on disk space.

7. Luise tries to create a Request for Remote statistics. When she tries to open the correct database to configure this option, she cannot find the database. What is the name of the correct database, and why might it not have been created?

   a. EVENTS4.NSF, the EVENT and/or REPORT tasks are not loaded.

   b. EVENTS4.NSF, the STATS task is not loaded.

   c. STATREP.NSF, the EVENT and/or REPORT tasks aren't loaded.

   d. STATREP.NSF, the STATS task is not loaded.

8. What command should Heidi use if she needs to start reporting on her server?

   a. LOAD REPORTER

   b. LOAD EVENT

   c. LOAD REPORT

   d. LOAD STATS

9. What is the first result of loading EVENT on a server?

   a. Create EVENTS4.NSF.

   b. Create STATREP.NSF.

   c. Start REPORTER.

   d. Start REPORT.

10. How can Julie cause Event Trouble Tickets to be created automatically?

    a. Use EVENTS4.NSF to configure Trouble Ticket creation at a particular event or threshold.

    b. Use STATREP.NSF to configure Trouble Ticket creation at a particular event or threshold.

    c. She cannot. They must be created manually in the STATREP.NSF database.

    d. She cannot. They must be created manually in the EVENTS4.NSF database.

11. Paula wants to monitor ACL changes to NAMES.NSF. How can she configure this?

    a. Use the ACL Monitor action button in the Action menu of NAMES.NSF.

    b. Use the ACL Monitor action button in STATREP.NSF.

    c. Use the ACL Monitor action button in EVENTS4.NSF.

    d. Use the ACL Monitor action button in DBA4.NSF.

12. Carin wants to view all the statistics for her server. How can she do this?

    a. Use the By Stat view of the NAB.

    b. Use the By Stat view of STATREP.NSF.

    c. Type SHOW STATS ALL at the remote console.

    d. Type SHOW STAT at the remote console.

Answers

1. d. Both

2. c. Open the Notes Log on her server and look at the Database Sizes view.

3. b. Notes Log . . . Miscellaneous Events view

4. c. LOG = LOG.NSF, 1,0,5,45000

5. a. Use a Log Analysis to search for the name Server2 in the Notes Log database.

6. b. Use the EVENTS4.NSF database to configure a Statistic Monitor to monitor disk space and an Event Monitor. Cause the Event monitor to mail a notification to her when the Statistic Monitor causes an Event.

7. a. EVENTS4.NSF, the EVENT and/or REPORT tasks are not loaded.

8. c. LOAD REPORT.

9. c. Create EVENTS4.NSF.

10. c. She cannot. They must be created manually in the STATREP.NSF database.

11. c. Use the ACL Monitor action button in EVENTS4.NSF.

12. d. Type SHOW STAT at the remote console.

Index

Boldface page numbers denote illustrations.

tree structure of hierarchy, 5, **6**
trouble tickets, 318, 319–320, **320**
TYPE, 57

units of measure, 40
UNIX, 2
UPDALL program, 63, 250
UPDATE_NO_FULLTEXT, 58
UPDATE_SUPPRESSION_LIMIT, 58
UPDATE_SUPPRESSION_TIME, 58
UPDATERS, 58
Usage by Date/User, 306, **308**
users/user IDs/user names, 81, 82,
   83–85
  Access Control Lists (ACL),
    92–99, **92**
  anti-spoofing mechanism, 85
  authentication, 86–88, **87**
  denying access, Deny Access
    groups, 89
  grouping users, 260–264
  passwords, 31, 38, 84–85, **84**
  Person documents, NAB, 51–53,
    52
  recertifying users or servers,
    216–218, **216–219,** 220
  registering users, 27–32, **27–30,**
    32
  registration batch file, 30–31
  renaming a user, 214–215, **214,**
    215, 220
  Restrictions, Server document,
    NAB, 67–68, **67,** 88–90, **89**
  roles for users, 85, 92, 102–103,
    103

SHOW USERS console
  command, 229, **229**
text files, delimited text
  formats, 30
Usage by Date/User, 306, **308**
User IDs, 28–29, 38
User profiles, User Setup Profile
  documents, 49, 70–71, **71,**
  239–240, **239**
user unique organizational unit
  (UUOU), 10, 10, 29
USER.ID file, 20
user interface/GUIs, 2, 40, **40**
user preferences options, 40, **40**
User Setup Profile document, NAB,
  49, 70–71, **71,** 239–240, **239**
user unique organizational unit
  (UUOU), 10, **10,** 29
UserCreator role, access control,
  NAB, 105
UserModifer role, access control,
  NAB, 105

View Access Lists, 85, 99–101,
  100, 195

warning (high) events, 315
warning (low) events, 315
Warning Threshold, database
  size limit, 237
wide area networks (WANs), 14,
  59, 138, 179
winning domains, 293

X.500 naming standards, 5

About the Author

Libby Ingrassia Schwarz is a Certified Lotus Professional for both Application Development and System Administration. She is also a Microsoft Certified Systems Engineer specializing in Windows NT, TCP/IP, and Exchange. Her training credentials include being a Certified Lotus Instructor, Microsoft Certified Trainer, and Certified Technical Trainer.

Libby has been in the computer industry doing technical writing, training, course development, and consulting in Notes, NT, and other technologies since 1994. She has planned and implemented installations, taught classes, and developed applications for a variety of technologies. She currently writes and teaches for Total Seminars, LLC. You can send her e-mail at LibbyS@TotalSem.com.

Total Seminars is a training company specializing in government and corporate on-site training and training material development. Total Seminars instructors have written books and manuals on A+ Certification, Networking, MCSE Certification, and Lotus Notes. For information on books and on-site training courses, see their Web page at www.totalsem.com.

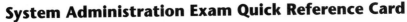

System Administration Exam Quick Reference Card

Here are some key study points to remember when taking the exams, even if you do not remember anything else. They should be the last items you run through before taking your exam. They are not the only pieces of information necessary to pass, but they give you some basic points to memorize.

Lotus Notes System Administration I: Exam 190-274

INSTALLATION AND CONFIGURATION

Organizations are used for security, while domains are used for mail routing.

The two main requirements for an NNN are constant connection and same protocol. The default NNN is Network1.

The Organization certifier is contained in the CERT.ID.

The elements of a hierarchical organization are: country codes (optional), organizations (required), organizational units (up to four; optional), and a common name (required).

Fully distinguished names can be canonical (CN=Libby Schwarz/OU=Instructors/O=TotalSem) or abbreviated (Libby Schwarz/Instructors/TotalSem).

Use the Server Administration panel to register new users, servers, and certifiers.

Notes creates CERT.ID, SERVER.ID, NAMES.NSF, and a USER.ID when the first server in an organization is configured.

THE NAME AND ADDRESS BOOK

One domain is related to one Public NAB.

The NAB contains the following types of documents: Group documents, Person documents, Location documents, Certificate documents, Server Configuration documents, Server Connection documents, Domain, Mail-In database documents, Server Programs documents, Server documents, and User Setup Profile documents.

SECURITY

Time-delay features of ID files prevent password-guessing programs.

Anti-spoofing features of ID files (hieroglyphics) prevent password-capturing programs.

Users must authenticate with servers to gain access. Authentication is a two-way process that uses the public-private keys in the user ID files. If you do not have a certificate in common, you cannot authenticate.

Users can be denied access to servers using the Restrictions section of the Server document.

The ACL has seven levels of access: Manager, Designer, Editor, Author, Reader, Depositor, and No Access.

MAIL ROUTING

The router is the server-based task that transfers and delivers mail. The mailer is an element of the workstation software that moves your outgoing messages to your server's MAIL.BOX.

Mail routes automatically and immediately between servers that are in the same NNN.

Use Connection documents to schedule replication and mail routing outside your NNN.

A routing threshold is the number of messages your router waits for before routing outside the schedule.

The default cost for routing across a LAN is 1; across a dial-up connection, it is 5.

Message-based mail is the default type of mail. To enable Shared mail, use the TELL ROUTER USE command.

Shared mail is used to save space on a server.

When you enable Shared mail, all new messages automatically use Shared mail. You do not have to do anything else. If you want messages sent before you enabled Shared mail to be stored in the single copy object store, use the LOAD OBJECT LINK command.

If you save sent mail, encrypt incoming mail, or edit mail you have received, those messages are not stored in the single copy object store.

If SHARED_MAIL=1, Shared mail is used for delivery only; however, only messages sent to more than one user will be stored in the single copy object store.

If SHARED_MAIL=2, Shared mail is used for delivery and transfer; however, messages sent to any number of users will be stored in the single copy object store.

If you need to move a mail file that uses Shared mail, the first thing you must do is unlink it from the current Shared mail file.

If you are having problems with mail, use the MAIL.BOX to find and release dead and pending mail. You can also send a mail trace using File . . . Tools . . . Server Administration . . . Mail . . . Send Mail Trace.

REPLICATION

Replication in Notes 4 occurs at the field level.

If Merge is enabled, replication conflicts will not be created when users edit different fields on the same document in different replicas of the database.

The four replication types are pull only, push only, pull-pull, and pull-push.

Use Connection documents to schedule replication. If you need to force replication, use the PUSH, PULL, and REPLICATE commands at the console.

The default replication type is PULL-PUSH.

If deleted documents are reappearing during replication, your purge interval occurs more quickly than your replication schedule.

A server should have at least Editor access to replicate new and changed documents. A server can replicate new documents only with Author access.

You must edit the server document for Passthru servers to allow Passthru. If the Passthru restriction fields are blank, no one is allowed to use Passthru to access the server.

Notes v3 servers can only be the final destination of Passthru.

ADMINISTRATION TOOLS AND TASKS

Use AdminP to recertify or rename users and servers.

The following are the prerequisites to use AdminP: ADMIN4.NSF must exist, CERTLOG.NSF must exist, the NAB must have an Administration server designated, all servers involved must be r4 or later, and the environment must use hierarchical naming.

A database limit is set at database creation and cannot be changed. When a user tries to save a document when the database is over its limit, they receive an error and cannot save.

A database quota should be set below the limit and will give an error to the user who it reaches. A database quota can be changed.

A database threshold should be set below the quota and will not give an error to the user when it is reached. Instead, it will write a message to the log. A threshold can be changed.

If you are using Remote Passthru, your user must have a Passthru server designated in either a Location document or in a Connection document.

To trace Network connection problems, use File . . . Tools . . . User Preferences . . . Ports . . . Trace Connection. This sends a trace to the specified server or other location that logs Summary Progress Information by default.

Lotus Notes System Administration II: Exam 190-275

ADVANCED SETUP AND CONFIGURATION

You can have multiple organizations in a single domain and multiple domains in a single organization.

If you communicate with another hierarchical organization, you must cross-certify with that organization to authenticate.

Cross-certification can occur at any level of the organization: O, OU, or Server to O, OU, Server, or User.

When communicating with a non-hierarchical organization, you must obtain flat certificates from that organization.

To send mail outside your domain, use explicit naming, such as Libby/TotalSem @TotalSem @ACME, if you will use the ACME domain to reach the TotalSem domain.

An adjacent domain is physically connected to your domain. Create an adjacent domain document and a server Connection document to route mail.

A non-adjacent domain is not physically connected to your domain. You will use an adjacent domain to deliver your mail to the non-adjacent domain. Create a non-adjacent domain document, an adjacent domain document, and a server Connection document to route mail.

To stop routing mail from one domain to another, use the domain document to specify the domain from which to deny mail.

Use the NAMES= line in the NOTES.INI to cascade Name and Address Books from multiple domains.

SERVER MONITORING

Use the Notes Log and the Notes Log analysis to monitor your Notes servers.

Use EVENTS4.NSF (the Statistics and Events database) to configure events, statistics, thresholds, and reporting for your servers.

Use STATREP.NSF (the Statistics Reporting database) to view events, alarms, statistics, and reports for your servers.

REPORTER and EVENT are required tasks on the server to collect and report statistics and events.

Use LOAD REPORT or LOAD EVENT to start these tasks on the server. Add them to the SERVERTASKS= line of the NOTES.INI to have them run all the time.

Trouble tickets must be created manually in the Statistics and Reporting database (for events and alarms).

Use the database tools to analyze a database. The analysis will be stored locally in the Database Analysis database (DBA4.NSF) by default.